憶猶未盡之住家菜

A Gourmet Chef's
Home Cooking

劉冠麟 著

By Eddie Liu

肉類
Meat: Beef, Pork, Chicken and Duck

010 滷牛舌
Stewed Beef Tongue

012 蘿蔔炆牛腩
Stewed Beef Brisket with Radish

014 清燉黃牛肉
Beef Cheek Stew in Clear Broth

016 花生炆豬尾
Pork Tails and Peanuts Stew

018 南乳炆豬手煲
Braised Front Pork Legs in Red Pickled Bean Curd

020 甜梅菜扣肉
Steamed Pork with Sweet Pickled Mustard Greens

022 欖角豉汁蒸排骨
Steamed Ribs in Black Bean and Black Olive Sauce

024 鹹蝦蒸腩仔
Steamed Pork Belly in Shrimp Sauce

026 野菇紅燒肉
Braised Pork Belly with Wild Mushrooms in Soy Sauce

028 羊腩煲
Mutton Brisket Stew

030 生炆鵝掌
Braised Goose Webs

032 八寶填鴨
Steamed Duck with "Eight Treasures"

034 金針雲耳蒸雞
Steamd chicken with Enoki Mushrooms and Black Mushrooms

036 咖哩薯仔炆雞翼
Braised Chicken Wings and Potatoes in Curry Sauce

038 梅菜心蒸豬肉餅
Steamed Minced Pork Patty with Pickled Mustard Greens

040 薑汁豬腦煎蛋
Pan-fried Pork Brains with Eggs and Ginger

042 煎鑲藕餅
Fried Lotus Roots and Pork Cakes

044 瑤柱肉碎蒸水蛋
Steamed Eggs with Scallop and Ground Pork

海鮮類
Seafood:
Fish, Crab, Shrimp, Clams, Oysters and More

清蒸黃腳立 048
Steamed Black Snapper

薑蔥焗鯉魚 050
Braised Carp with Ginger and Scallions

生抽煎鯧魚 052
Pan-fried Pomfret in Soy Sauce

薑蔥焗魚頭 054
Braised Fish Heads with Ginger and Scallions

黃魚凍 056
Yellow Croaker Jelly

生抽煎大蝦碌 058
Pan-fried Prawns in Light Soy Sauce

梅菜蔥段蒸鯇魚腩 060
Steamed Grass Carp Belly with Pickled Mustard Greens and
Sectioned Scallions

陳皮薑米焗鯇魚腸 062
Steamed Grass Carp Intestines with Dried Tangerine Peels and
Ginger Rice

陳皮豉汁蒸白鱔 064
Steamed Eel with Dried Tangerine Peels and Black Bean Sauce

薑蔥焗花蟹 066
Braised Spotted Crabs with Ginger and Scallions

薑蔥焗生蠔 068
Braised Oysters with Ginger and Scallions

豉汁炒蜆 070
Stir-Fried Clams in Black Bean Sauce

鹹魚蓉鑲豆腐 072
Tofu stuffed with Minced Salted Fish Paste

鹹魚蒸肉餅 074
Steamed Minced Pork Patty with Salted Fish

鮮蠔汁鮑魚 076
Abalone in Oyster Sauce

蔬菜和蛋
Vegetalbes and Eggs

082 青瓜馬蹄炒圍蝦
Stir-fried Cucumbers and Water Chestnuts with Shrimps

084 蝦蜆肉炒韭菜花
Stir-fried Chive Flowers with Shrimps and Clams

086 勝瓜雲耳炒洋蔥
Zucchinis, black Mushrooms and Onions Stir Fry

088 蠔油涼瓜炒牛肉
Stir-fried Beef Fillets and Bitter Gourds in Oyster Sauce

090 鹹魚蓉鑲苦瓜煲
Bitter Gourds with Salted Fish Stuffing

092 薑汁鹹魚蓉炒芥蘭
Stir-fried Kailan with Ginger Juice and Minced Salted Fish

094 蘿蔔煮魚餅
Fish Cake and Radish

096 蜆肉菜脯煎蛋
Omelet with Dried Radishes and Clams

098 奄仔蟹蒸水蛋
Steamed Eggs with Crabs

100 番茄煮雞蛋
Sautéd Tomatoes and Eggs

102 鹹蛋皮蛋蒸肉餅水蛋
Steamed Pork Patty with Salted, Preserved and Fresh Eggs

煲湯和麵飯
Soup, Noodles and Rice

眉豆花生豬骨煲雞腳湯 106
Black-eyed Peas, Peanuts, Pork Bones and Chicken Feet Stew

花膠老雞煲响螺頭 108
Stewed Chicken with Fish Maws and Conches

青紅蘿蔔無花果煲牛坑腩 110
Stewed Beef Brisket with Green Radishes, Carrots and Figs

章魚蜜棗陳皮豬腱煲蓮藕 112
Stewed Pork Hock and Lotus Roots with Octopuses, Candied Dates and
Dried Tangerine Peels

南北杏菜乾煲豬肺 114
Stewed Pork Lungs with Sweet and Bitter Almonds and Dried Vegetables

八寶冬瓜盅 116
Stewed Wax Gourd with "Eight Treasures"

西洋菜陳腎煲龍骨 118
Stewed Pork Backbones and Roast Duck Kidneys with Watercress

生木瓜陳皮老薑煎魚湯 120
Stewed Fish with Green Papaya, Dried Tangerine Peels and Ginger

白胡椒老薑煲豬肚龍骨 122
Stewed Pork Stomachs and Backbones with White Pepper and Ginger

赤小豆粉葛沙參玉竹煲魚湯 124
Fish Soup with Red Beans, Acorus Gramineus, Sea Cucumbers and
Polygonatum Odoratum

豉油皇炒麵 126
Fried Egg Noodles in Soy Bean Sauce

水蟹粥 128
Crab Rice Porridge

基圍蝦田雞粥 130
Shrimp and Frog Rice Porridge

豬肉雞蛋豉油撈飯 132
Steamed Rice Mixed with Lard, Eggs and Soy Sauce

one
肉類

Meat: Beef, Pork,
Chicken and Duck

滷牛舌

過年過節、打牌、喝酒、看電視，滷牛舌是最好的零食，也可當作年菜。
牛舌滷得好真的是軟滑可口。

食材：
牛舌1條（約2公斤）
薑片100公克
花椒20公克
八角20公克
胡椒粒20公克
香菜50公克

調味料：
高粱酒100cc
清水3000cc
生抽2湯匙
老抽2湯匙
蠔油3湯匙
冰糖100公克

作法：
1. 牛舌用水煮15分鐘去雜質，去皮。
2. 把所有食材與調味料加入清水中，加蓋慢火燜煮2小時即可。

Stewed Beef Tongue

This is a snack that one may enjoy when playing majiang, drinking alcoholic beverages and watching TV during holidays. It can also be a part of the Spring Festival feast. Good stewed beef tongue is tender in texture and great in flavor.

Ingredients:
One beef tongue (about 2kg)
100g sliced ginger
20g Sichuan pepper
20g star anise
20g peppercorns
50g Chinese parsley

Seasonings:
100ml Kaoliang spirit
3000ml water
2 tbsp light soy sauce
2 tbsp dark soy sauce
3 tbsp oyster sauce
100g rock sugar

Directions:
1. Put beef tongue in boiling water and cook for 15 minutes to remove the coating.
2. Add all the ingredients and seasonings in water and put a lid on the pot. Cook at a lower temperature for 2 hours.

Stewed Beef Brisket with Radish

蘿蔔炆牛腩

蘿蔔炆牛腩煲是一道香氣十足的下酒好菜。
牛坑腩是香港人的叫法，指的是牛肋骨部位，也就是台灣人說的牛腩。
蘿蔔鮮甜，牛腩香嫩，野味十足。

食材：
牛坑腩1公斤
白蘿蔔1公斤
老薑50公克
八角4粒
胡椒粒20粒
高湯2000cc

調味料：
生抽1湯匙
老抽1湯匙
白糖1湯匙

作法：
1. 將牛腩先煮15分鐘去雜質，撈起切塊備用。
2. 砂鍋加入高湯煮滾，加入香料、調味料、牛腩，煮約90分鐘。
3. 加入切塊的白蘿蔔，再煮半小時收汁即可。

Stewed Beef Brisket with Radish

This is a dish of aroma and goes very well with alcoholic beverages. Beef brisket, is a cut of meat in the ribs. It is called niunan in Taiwanese. The radish is flavorful, and the beef is tender in texture. The combination offers a taste of wildness.

Ingredients:
1kg beef brisket
1kg Chinese radish
50g ginger
4 star anise
20 peppercorns
2000ml soup stock

Seasonings:
1 tbsp light soy sauce
1 tbsp dark soy sauce
1 tbsp sugar

Directions:
1. Blanch beef brisket in boiling water for 15 minutes. Drain well. Cut the beef into cubes.
2. Heat soup stock in a casserole. Add the spices, seasonings and beef cubes and cook for 90 minutes.
3. Add chunked radishes and cook for 30 more minutes or until the broth thickened.

清燉黃牛肉

這道菜的湯頭清甜，臉頰肉滑嫩。

食材：
牛臉頰肉1公斤
洋蔥2個
牛肋骨1公斤
老薑100公克
清水1000cc

調味料：
鹽少許

作法：
1. 將牛頰和牛肋骨汆燙去血水。
2. 放入大燉盅，加入洋蔥、老薑、清水，燉3小時，取肉切塊，湯加鹽調味即可。

Beef Cheek Stew in Clear Broth

The broth is fresh and sweet, and the beef tastes smooth and soft.

Ingredients:
1kg beef cheeks
2 onions
1kg beef ribs
100g ginger
1000ml water

Seasonings:
A bit salt

Directions:
1. Blanch the beef.
2. Put the blanched beef in a large stewing pot. Add onions, ginger, water and cook for three hours. Take out the beef and cut it into cubes. Add a bit salt to the broth.

Beef Cheek Stew in Clear Broth

花生炆豬尾

這是廣東人的年菜，多在冬天食用，也是一般家庭都會做的家常菜。
廣東人以花生炆豬尾補腎補腰。豬尾鮮嫩含膠質，煮到入花生裏，那種香氣令我抵擋不了。

食材：
豬尾巴6條（約800公克）
大粒帶皮花生300公克
高湯2000cc
薑片100公克

調味料：
柱侯醬3湯匙
白糖1茶匙
生抽2茶匙
料理酒2湯匙

作法：
1. 豬尾巴洗淨拔毛。
2. 加入花生、調味料一同放入高壓快鍋煮約50分鐘，再放入砂鍋收汁，勾芡即可。

Pork Tails and Peanuts Stew

This is a dish that the Cantonese may savor during the Chinese New Year holidays. It is usually made in winter and is very popular among the ordinary households. The Cantonese believe that this dish is good for kidneys. Pork tails are tender in texture and rich in colloid. When the broth is absorbed by the peanuts, the aroma is irresistible.

Ingredients:
6 pork tails (about 800g)
300g peanuts with skins on
2000ml soup stock
100g sliced ginger

Seasonings:
3 tbsp Chu Hou bean paste
1 tsp sugar
2 tsp light soy sauce
2 tbsp cooking alcohols

Directions:
1. Clean the pork tails and remove all the bristles.
2. Add pork tails, peanuts and seasonings in a pressure cooker and cook for around 50 minutes. Then move them into a casserole and cook until the broth is thickened. Add potato starch water to give it the last touch.

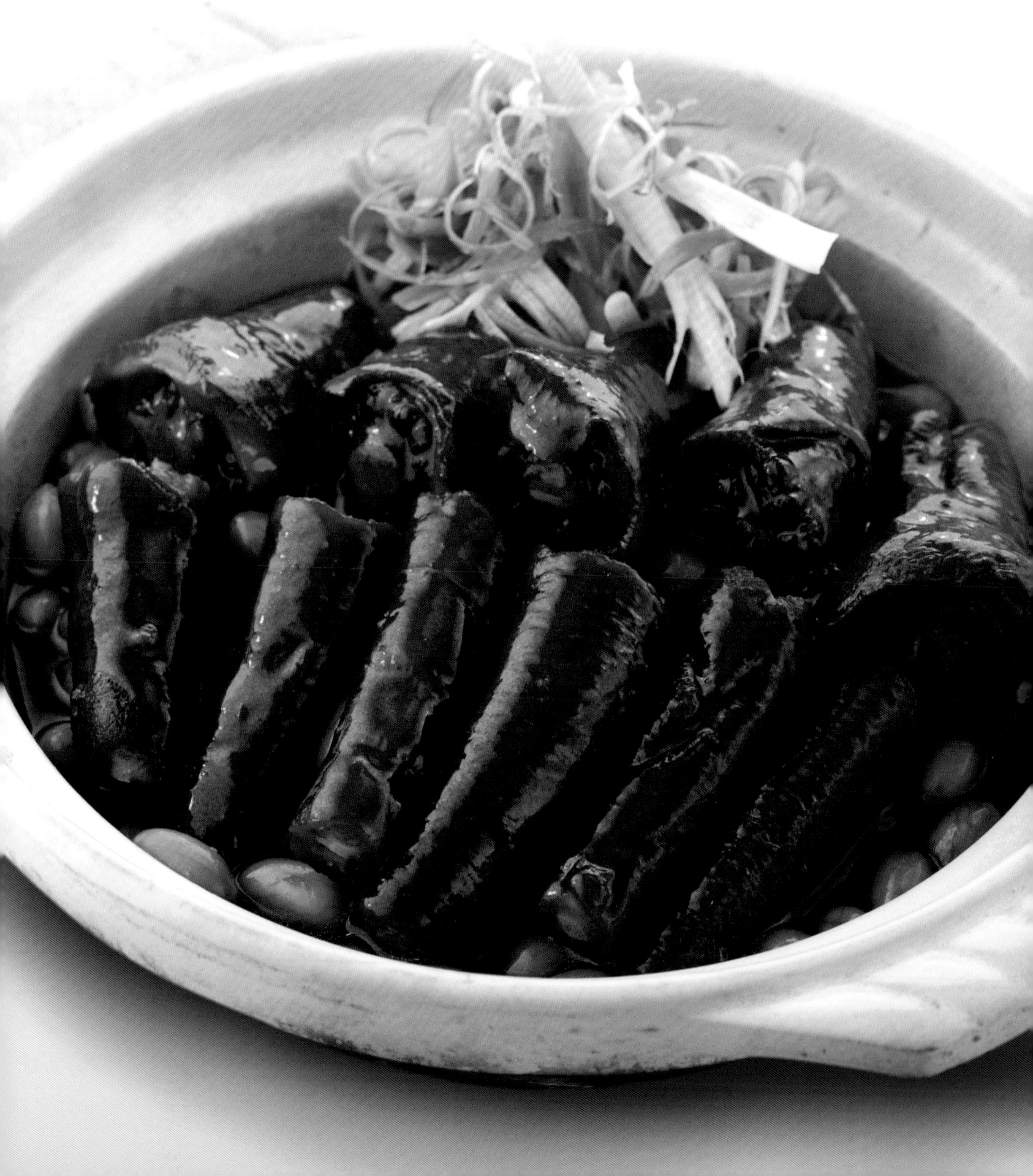

Pork Tails and Peanuts Stew

南乳炆豬手煲

上海人和廣東人都喜歡用南乳，即紅豆腐乳，去燜燒食物，排骨或豬手。
南乳炆豬手是順德名菜，Q彈可口，香氣十足。

食材：
前豬腳1對
薑片50公克
蒜苗

調味料：
生抽1茶匙
白糖1湯匙
高湯2000cc
廣東紅腐乳500公克（打成漿）

作法：
1. 豬腳切開約5公分大小，毛要拔乾淨，汆燙去血水。
2. 放入砂鍋內，加入調味料，慢火煲煮90分鐘收汁即可。

Braised Front Pork Legs in Red Pickled Bean Curd

People in Shanghai and Guangdong like to use nanru, or red pickled bean curd, to braise ribs or front pork legs. This is a well-known Shuande delicacy. It is springy in texture and smells really good.

Ingredients:
A few front pork legs
50g sliced ginger
Shredded leeks

Seasonings:
1 tsp light soy sauce
1 tbsp sugar
2000ml soup stock
500g mashed Cantonese-style red pickled bean curd

Directions:
1. Chop the front pork legs into 2-inch chunks. Clean the bristles thoroughly and blanch the pork legs.
2. Put the pork legs chunks in a casserole and add seasonings. Cook at a low temperature for 90 minutes or until the broth sickened.

Steamed Pork with Sweet Pickled Mustard Greens

甜梅菜扣肉

我相信很多朋友都會喜歡客家甜梅菜扣肉。
我姊夫是客家人，他教我做很多客家菜；我覺得甜梅菜和扣肉真的很對味。

食材：
甜梅菜心300公克
帶皮五花肉600公克
蒜苗100公克

調味料：
生抽1茶匙
老抽2茶匙
蠔油2茶匙
麻油1茶匙
白糖2茶匙

作法：
1. 梅菜洗淨切碎。
2. 帶皮五花肉用少許老抽上色，炸至金黃後切片，約中指寬度大小。
3. 將所有調味料拌好。
4. 將五花肉先放入盆內，再加入調味料與梅菜、蒜苗，蒸3小時即可。

Steamed Pork with Sweet Pickled Mustard Greens

I believe that many people like this Hakka delicacy. My brother-in-law is a Hakka, and I have learned many Hakka cuisines from him. Sweet pickled mustard greens, in my opinion, are a great fit with pork.

Ingredients:
300g sweet pickled mustard greens
600g pork belly with skin
100g shredded leaks

Seasonings:
1 tsp light soy sauce
2 tsp dark soy sauce
2 tsp oyster sauce
1 tsp sesame oil
2 tsp sugar

Directions:
1. Rinse mustard greens and then mince.
2. Marinate pork belly in dark soy sauce. Fry until golden brown and cut into ½ inch slices.
3. Mix all the seasonings.
4. Place sliced pork belly in a bowl. Add seasonings, minced mustard greens and shredded leaks and steam for 3 hours.

Steamed Ribs in Black Bean and
Black Olive Sauce

欖角豉汁蒸排骨

潮洲欖角醒胃，加上豉汁蒸出來的肉或魚，都是極品。

食材：
潮州欖角20公克
黑豆豉20公克
蒜茸20公克
紅椒10公克
豬腩排500公克

調味料：
生抽1茶匙
蠔油1茶匙
白糖茶1匙
太白粉1茶匙

作法：
1. 潮州醃製橄欖角切碎、黑豆豉切碎。
2. 起油鍋，少許油將橄欖碎及豆豉加白糖炒香。蒜茸另外炒至金黃色。
3. 排骨斬成大拇指大小，所有材料及調味料拌勻，蒸30分鐘即可。

Steamed Ribs in Black Bean and Black Olive Sauce

Black olives are good for stomach. The taste is phenomenal when they are cooked with meat or fish in black bean sauce.

Ingredients:
20g pickled black olives from Chaozhou
20g fermented black beans
20g finely minced garlic
10g red chili peppers
500g pork spareribs

Seasonings:
1 tsp light soy sauce
1 tsp oyster sauce
1 tsp sugar
1 tsp potato starch

Directions:
1. Mince pickled black olives and fermented black beans.
2. Heat a bit of oil in a wok. Add the minced olives, black beans and sugar and stir-fry until fragrant. Fry minced garlic in another wok until golden brown.
3. Chop the ribs into 2-inch-long pieces. Mix all the ingredients and seasonings evenly and steam for 30 minutes.

鹹蝦蒸腩仔

香港人除了喜歡鹹魚之外，還喜歡蝦醬。
但不是每個人都可以接受蝦醬，能吃蝦醬蒸五花肉的，一定是上等人，
因為怎麼難聞他們都可以接受。哈哈哈，鮮鮮鮮。

食材：
蝦醬50公克
五花肉400公克
薑絲100公克

調味料：
太白粉1茶匙
白糖2茶匙
米酒2茶匙
油1茶匙

作法：
1. 五花肉切片。
2. 將所有材料與調味料拌好，放入盤中蒸30分鐘即可。

Steamed Pork Belly in Shrimp Sauce

In addition to salted fish, many people in Hong Kong like shrimp sauce as well.
However, not everyone can enjoy shrimp sauce. Those who dare to try this dish must be
the toughest because it smells really fishy. Oh, how I love its fishiness.

Ingredients:
50 shrimp sauce
400g pork belly
100g shredded ginger

Seasonings:
1 tsp potato starch
2 tsp sugar
2 tsp rice wine
1 tsp oil

Directions:
1. Slice the pork belly.
2. Mix all the ingredients and seasonings, place the meat in a plate and steam for 30 minutes.

Steamed Pork Belly in Shrimp Sauce

野菇紅燒肉

食材：
五花肉600公克
蘑菇300公克

調味料：
淡味老抽 30公克
鹹味生抽 10公克
冰糖100公克
雞粉10公克
上湯500公克

作法：
1. 五花肉切成大拇指大小。
2. 起油鍋，放入蔥、薑和五花肉，炒香，再放入所有調味料。
3. 上湯以小火燜煮35分鐘，汁收乾即可。

Braised Pork Belly with Wild Mushrooms in Soy Sauce

Ingredients:
600g pork belly
300g mushrooms

Seasonings:
30ml dark soy sauce
10ml light soy sauce
100g rock sugar
10g chicken powder
500ml stock

Directions:
1. Cut the pork belly into thumb-sized pieces.
2. Heat a wok. Add scallions, ginger and the pork belly, and Stir-fry until fragrant. Add all of the seasonings.
3. Put a lid on the wok and cook over low heat for another 35 minutes or until the broth thickens.

羊腩煲

冬天時，香港的廣東人喜歡拿羊腩煲做為一道主菜，以明爐炭火加上砂鍋，
一面吃羊肉，一面燙青菜，青菜以唐生菜和茼蒿菜為主。
台灣羊肉爐以藥材湯多，香港的羊腩煲不放藥材，以汁多，是路邊大排檔的名菜。

食材：

帶骨羊腩1000公克	白腐乳50公克
支竹400公克	南乳50公克
馬蹄100公克	柱侯醬100公克
冬菇100公克	甘蔗600公克
冬筍100公克	清水2000cc
老薑100公克	

作法：
1. 羊腩用火烤，至皮呈金黃色。
2. 砂鍋洗淨備用，支竹用油泡發備用，馬蹄去皮備用。
3. 起油鍋，將羊腩肉炒乾，放入所有食材、清水，慢火煮1.5小時，試調味即可。

Mutton Brisket Stew

This is a main course that the Cantonese people in Hong Kong like to make in winter. They cook mutton in a casserole over a charcoal burner and scald green vegetables such as Chinese lettuce or garland chrysanthemum in the stock. Most of the Taiwanese-style mutton stews are cooked with Chinese herbs. In Hong Kong, however, we do not add herbs in this famous street food stall delicacy.

Ingredients:

1kg mutton brisket	50g white pickled bean curd
400g tofu skin	50g red pickled bean curd
100g water chestnut	100g chu hou bean paste
100g dried mushrooms	600g sugar canes
100g bamboo shoots	2000ml water
100g ginger	

Directions:
1. Roast mutton brisket until the skin turns golden brown.
2. Rinse a casserole. Soak dried tofu skin in oil. Skin water chestnuts.
3. Heat oil in a wok. Fry mutton until crisp. Add all the ingredients and water.
   Cook at a low temperature for 1.5 hours. Add a bit salt and serve.

Mutton Brisket Stew

生炆鵝掌

生炆鵝掌是香港酒席菜宴上的名菜。
鵝掌先炸後炆，吃在嘴裏軟滑香嫩，讓人連手指都想吞下去，我願用我的鮑魚來換你的鵝掌。

食材：
鵝掌10對（20只）
火腿200公克
薑片100公克
蔥段100公克

調味料：
雞湯600cc
冰糖30公克
蠔油100cc
生抽50cc

作法：
1. 鵝掌先用老抽上點色，再下油鍋炸一下，約1分鐘撈起備用。
2. 放入快鍋。
3. 加入所有食材及調味料，約1小時開蓋，取出鵝掌放入砂鍋收汁，勾芡即可。

Braised Goose Webs

This is a very famous banquet cuisine in Hong Kong. The goose webs, after fried and braised, are so smooth and tender that they almost melt in one's mouth. I would rather have these goose webs than abalones.

Ingredients:
20 goose webs
200g ham
100g sliced ginger
100g sectioned scallions,
    aka green onions

Seasonings:
600ml chicken stock
30g rock sugar
100ml oyster sauce
50ml light soy sauce

Directions:
1. Marinate goose webs in dark soy sauce. Fry them in boiling oil for one minute.
2. Put them in a pressure cooker.
3. Add all the other ingredients and seasonings and cook for one hour. Remove them from the cooker and put them in a casserole. Cook until the sauce is thickened and add potato starch water to give the last touch.

Steamed Duck with
"Eight Treasures"

八寶填鴨

八寶填鴨是非常繁複的做功菜，所以每回我想吃就只能跟我奶媽預訂。
直到現在，我都還會想吃呢！

食材：
光鴨1隻（約1500公克）
冬菇50公克
炸魚肚100公克
水發魷魚100公克
糯米300公克
薑片20公克
蔥段20公克

調味料：
生抽少許
蠔油少許
白糖少許

作法：
1. 將光鴨洗淨去內臟，從背敲鬆取出鴨骨。
2. 用熱水汆燙至老抽顏色，再炸至金黃色，備用。
3. 所有的食材（鴨子除外）放入熱水汆燙後，用薑蔥爆香，加入蠔油、生抽炆約10分鐘，塞放入鴨肚裡。
4. 蒸80分鐘，放入油鍋炸至金黃色，上盤子前勾薄油芡便可。

Steamed Duck with "Eight Treasures"

This delicacy requires high cooking skills. My nanny used to make it for me if I ordered in advance. It is the food that I have always craved for.

Ingredients:
1 duck (about 1500g)
50g dried mushroom
100g fried fish bellies
100g soaked squid
300g glutinous rice
20g sliced ginger
20g sectioned scallions

Seasonings:
A bit soy sauce
A bit oyster sauce
A bit sugar

Directions:
1. Clean the duck and remove entrails. Debone from the back.
2. Blanch the duck in boiling water. Then deep fry until golden brown. Set aside.
3. Blanch all the ingredients except the duck in boiling water. Stir-fry ginger and scallions until fragrant. Add oyster sauce and light soy sauce and cook at low temperature for around 10 minutes. Stuff the cavity of the duck with the stuffing.
4. Steam for 80 minutes and deep fry until golden. Serve with a bit of oil.

Steamd chicken with Enoki Mushrooms and Black Mushrooms

金針雲耳蒸雞

金針雲耳蒸雞是我最喜歡的下飯菜。
金針的鮮、雲耳的脆、雞肉的嫩，醬香加上多汁，用來拌飯很正點。

食材：
雞腿肉500公克
乾金針50公克
水發雲耳100公克
老薑片50公克
蔥段50公克

調味料：
醬油1茶匙
砂糖1茶匙
麻油3滴
太白粉少許
鹽1/2茶匙

作法：
1. 雞腿肉洗淨斬件。
2. 將所有食材放入深盤中蒸20分鐘即可。

Steamd chicken with Enoki Mushrooms and Black Mushrooms

This is my favorite dish to go with rice. The freshness of enoki mushrooms, the crispness of black mushrooms and the tenderness of chicken are mixed with the fragrance of the sauce. It goes with rice very well.

Ingredients:
500g chicken legs
50g dried enoki mushrooms
100g soaked black mushrooms
50g sliced ginger
50g sectioned scallions

Seasonings:
1 tsp soy sauce
1 tsp sugar
3 drops sesame oil
A bit potato starch
1/2 tsp salt

Directions:
1. Rinse the chicken legs and chop into parts.
2. Put all the ingredients and seasonings in a deep plate and steam for 20 minutes.

咖哩薯仔炆雞翼

咖哩薯仔炆雞翼只要炆煮個30分鐘就能馬上拿來配飯,是帶便當最好的選擇。

食材:
雞中翅10只
馬鈴薯500公克
高湯800公克
洋蔥50公克
薑茸20公克

調味料:
帶鹽牛油1茶匙
咖哩粉3茶匙
辣椒粉1茶匙
辣椒油1茶匙

作法:
1. 雞翼拉油至金黃色備用。
2. 馬鈴薯切成大塊,油炸至金黃備用。
3. 起油鍋,少許沙拉油加入洋蔥、薑茸,爆香後放入所有調味料爆香。
4. 砂鍋加入雞翼、馬鈴薯、高湯,炆煮30分鐘即可。

Braised Chicken Wings and Potatoes in Curry Sauce

It takes only 30 minutes to make the dish. It goes well with rice and is perfect for lunch box.

Ingredients:
10 chicken lower wings
500g potatoes
800ml soup stock
50g onions
20g finely minced ginger

Seasonings:
1 tsp salted butter
3 tsp curry powder
1 tsp chili powder
1 tsp chili oil

Directions:
1. Fried the wings until golden brown.
2. Chunk the potatoes. Deep fry until golden brown.
3. Heat a bit of vegetable oil in a wok. Add onions, ginger and stir-fry until fragrant. Add all the seasonings and continue to toss until fragrant.
4. Put chicken wings, potatoes and soup stock in a casserole and cook at a low temperature for 30 minutes.

梅菜心蒸豬肉餅

清蒸的菜餚一向比較清火，夏天配飯的梅菜香肉之嫩，令人開胃。

食材：
甜梅菜心200公克
豬梅花絞肉300公克
豬五花絞肉150公克
薑絲50公克

調味料：
生抽50cc
蠔油80cc
白糖50公克
太白粉1茶匙

作法：
1. 將梅菜心洗淨切碎，放入熱鍋內炒乾備用。
2. 梅花肉與五花肉加入所有調味料捶打拌好。
3. 加入梅菜，擺在盆上，加入薑絲，蒸20分鐘即可。

Steamed Minced Pork Patty with Pickled Mustard Greens

The traditional Chinese medicine believes that steamed food tend to reduce the internal body heat. The patty is tender in texture which helps eat more rice in summer.

Ingredients:
200g sweet pickled mustard greens
300g ground pork collar-butt
150g ground pork belly
50g shredded ginger

Seasonings:
50ml light soy sauce
80ml oyster sauce
50g sugar
1 tsp potato starch

Directions:
1. Rinse the pickled mustard greens and then mince. Stir-fry the minced mustard greens in a hot wok until dried and crisp and then set aside.
2. Beat and mix pork collar-butt, pork belly and all seasonings.
3. Add pickled mustard greens and put it in a bowl. Add shredded ginger and steam for 20 minutes.

*Steamed Minced Pork Patty with
Pickled Mustard Greens*

Pan-fried Pork Brains with Eggs and Ginger

薑汁豬腦煎蛋

中國人說吃腦補腦。
小時候，母親就怕我笨，一星期起碼讓我吃兩、三次的豬腦，不管是煎的還是燉的。

食材：
豬腦1副
生雞蛋3顆
薑米10公克

調味料：
胡椒粉1/3茶匙
白糖1/2茶匙
精鹽1茶匙
麻油1茶匙

作法：
1. 先將豬腦去紅色的血塊，用清水沖乾淨，煎至金黃色備用。
2. 蛋打散，加入薑米，再加入切成碎的豬腦，煎至焦香即可。

Pan-fried Pork Brains with Eggs and Ginger

The Chinese people believe you are what you eat. When I was little, my mom made me eat fried or stewed pork brains two or three times a week.

Ingredients:
1 pork brain
3 eggs
10g minced ginger

Seasonings:
1/3 tsp pepper power
1/2 tsp sugar
1 tsp salt
1 tsp sesame oil

Directions:
1. Remove the clots from the pork brain and rinse well. Fry until golden brown and set aside.
2. Beat eggs and add minced ginger and fried brain. Fry the mixture until brown and fragrant.

Fried Lotus Roots and Pork Cakes

煎鑲藕餅

煎釀蓮藕餅是順德名菜。煎好的蓮藕肉餅沾上少許蠔油，是配酒下飯的好菜。

食材：
蓮藕600公克
五花絞肉400公克
薑茸5公克
蔥茸5公克

調味料：
生抽2茶匙
蠔油2茶匙
白糖1茶匙
精鹽1/2茶匙
胡椒粉1/2茶匙
太白粉1½茶匙

作法：
1. 蓮藕洗淨切片，蒸熟後切碎。
2. 加入五花絞肉、薑茸、蔥茸及所有調味料，攪拌成圓球狀。
3. 將圓球壓扁，煎至金黃即可。可沾蠔油食用。

Fried Lotus Roots and Pork Cakes

This is a well-known Shunde cuisine. Fried Lotus Root Cake with oyster sauce goes very well with alcoholic beverages and steamed rice.

Ingredients:
600g lotus roots
400g ground pork belly
5g finely minced ginger
5g finely minced scallions

Seasonings:
2 tsp light soy sauce
2 tsp oyster sauce
1 tsp sugar
1/2 tsp salt
1/2 tsp pepper powder
1½ tsp potato starch

Directions:
1. Rinse the lotus roots and slice. Steam the root slices until soft and chop them into fine pieces.
2. Add ground pork, ginger, scallions and all the seasonings. Mix and roll the paste into balls.
3. Press flat each ball and pan-fry until golden brown. Serve with oyster sauce.

瑤柱肉碎蒸水蛋

家裡有不愛吃飯的小朋友，吃一餐飯可能要餵上一、兩個小時，
這種飲食習慣對健康不大好。所以，當媽媽與阿姨們就想到一個辦法，
用瑤柱肉去蒸蛋，拌菜飯給小朋友吃，營養又快速。

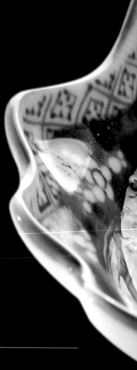

食材：
干貝50公克
豬絞肉100公克
雞蛋4顆
清水4顆雞蛋的量

調味料：
生抽1茶匙
鹽1/3茶匙
麻油2滴
太白粉1/2茶匙

作法：
1. 干貝用熱水泡發備用，豬絞肉加入麻油、太白粉攪拌均勻。
2. 雞蛋加入清水打散後即成蛋液，清水可用干貝水代替。
3. 蛋液加入豬絞肉、干貝打在一起，慢火蒸15分鐘即可，每5分鐘疏氣。

Steamed Eggs with Scallop and Ground Pork

It usually takes around one or two hours for a picky child to finish a meal. It is not a
healthy dining habit. That is why smart Chinese mothers and nannies come up with the
idea to make this dish, which is nutritious and easy to eat with steamed rice and
vegetables.

Ingredients:
50g dried scallops
100g ground pork
4 eggs
Water (equivalent to 4 eggs)

Seasonings:
1 tsp light soy sauce
1/3 tsp salt
2 drops sesame oil
1/2 tsp potato starch

Directions:
1. Soak dried scallops in hot water and set aside. Mix ground pork with sesame oil and potato starch evenly.
2. Add water in eggs and beat well. The water used to soak the dried scallops can be an option.
3. Mix the beaten eggs with ground pork and scallops and steam at a lower temperature for 15 minutes.
   Remove the cover every 5 minutes.

two
海鮮

Seafood: Fish, Crab, Shrimp, Clams, Oysters and More

清蒸黃腳立

港仔最愛的蒸魚也是我的最愛。為什麼要用黃腳立,而不用老鼠斑蘇眉?
因為黃腳立是比較平價的魚。野生的黃腳立生長在鹹水和淡水交界處的淡水中,
肉質鮮嫩,有一種特殊的魚油香氣,但是要買野生的不容易。

食材:
黃腳立或活黑格1條
（約600公克左右,黃腳立的鰭是黃色的）
薑片50公克
蔥絲100公克

調味料:
生抽3茶匙
魚露3茶匙
白糖2茶匙
（以上拌勻煮開,即成魚露醬油）
沙拉油2湯匙

作法:
1. 將魚洗淨,放入瓷盤中,擺上薑片。
2. 待水煮開,加蓋蒸10分鐘後,倒出魚水,拿掉薑片。
3. 所有調味料放入小鍋煮開備用。
4. 放上蔥絲,將沙拉油燒熱淋上,加入魚露醬油即可。

Steamed Black Snapper

Many Hong Kong people like steamed fish, and so do I. Why not use king garoupa or sumei rockfish to make the dish but black snapper? Black snappers are comparatively less expensive. Wild black snappers live at places where rivers meet the sea. They are tender in texture and their fat has a very unique aroma and flavor. However, it is not so easy to get wild black snappers.

Ingredients:
1 live black snapper or black porgy
 (about 600g)
50g sliced ginger
100g shredded scallions

Seasonings:
3 tsp light soy sauce
3 tsp fish sauce
2 tsp sugar
(Boil the above seasonings to make fish-flavored soy sauce.)
2 tbsp vegetable oil

Directions:
1. Rinse the fish and put it on a porcelain plate. Place sliced ginger on top of the fish.
2. Turn a steamer on to a hard boil. Put the fish in the steamer and cover. Cook for 10 minutes. Then lift the lid and remove the fish stock and ginger.
3. Put all the seasonings in a sauce pan and cook to boil.
4. Put shredded scallions on top of the fish and pour boiling vegetable oil over the fish. Add the fish-flavored soy sauce.

Steamed Black Snapper

薑蔥焗鯉魚

這是一道老式的廣東料理，大部分用砂鍋加上炭爐慢火燜煮。
煮出來的魚肉細嫩，魚卵香脆，是一道非常下飯的煲仔菜。

食材：
活鯉魚1條（約600公克）
薑片300公克
蔥段300公克

調味料：
生抽2茶匙
老抽2茶匙
甜麵醬1湯匙
蠔油1湯匙
白糖1茶匙

作法：
1. 將魚打鱗洗淨，魚腹剖開取腸泥，吸乾水備用。
2. 起鍋加入沙拉油，燒熱至80℃。
3. 放入鯉魚炸至金黃食撈起，將炸油倒出。
4. 起油鍋，加入薑蔥爆香。
5. 放入鯉魚，加入所有調味料及500CC的水，加蓋煮20分鐘，待醬汁收乾即可。

Braised Carp with Ginger and Scallions

This is a traditional Cantonese dish. A carp is usually braised in a casserole over a charcoal burner, which gives a delicate taste to the fish and a crisp texture to its roe. This clay-pot dish goes very well with steamed rice.

Ingredients:
1 live carp (about 600g)
300g sliced ginger
300g sectioned scallions

Seasonings:
2 tsp light soy sauce
2 tsp dark soy sauce
1 tbsp wheat paste
1 tbsp oyster sauce
1 tsp sugar

Directions:
1. Scale carp. Remove the entrails. Wipe the fish dry with a towel.
2. Heat vegetable oil in a wok to 80℃.
3. Fry the carp until it turns golden brown and pour away excess oil.
4. Heat oil in a wok. Add ginger and scallions and toss until fragrant.
5. Add the fish, all the seasonings and 500 ml of water. Put a lid on the wok and continue to cook for 20 minutes or until sauce thickened.

Braised Carp with Ginger and Scallions

生抽煎鯧魚

鯧魚在香港的潮州人口中叫鱠魚。在香港，大英倉很貴的，
大約要兩、三千元左右一尾。通常用來半煎炸，再用生抽、糖起鍋，增加香氣與味道。

食材：
白鯧魚600公克

調味料：
生抽3茶匙
白糖2茶匙

作法：
1. 鯧魚切成約4片，用布吸乾水分，拍上少許太白粉。
2. 起油鍋，煎至兩面酥脆。
3. 加入生抽及白糖，煎煮1分鐘即可。

Pan-fried Pomfret in Soy Sauce

Pomfret is called changyu in Teochiew. In Hong Kong, promfets are very expensive fish. It costs around NT$2,000-3,000 to get a pomfret. We usually pan-fry it and enhance its flavor with light soy sauce and sugar.

Ingredients:
600g white pomfret

Seasonings:
3 tsp light soy sauce
2 tsp sugar

Directions:
1. Cut the pomfret into four pieces. Dry them up with a towel. Dust a thin layer of potato starch over the fish steaks.
2. Heat oil in a wok and pan-fry the fish steaks until both sides turn crisp.
3. Add light soy sauce and sugar and continue to cook for one minute.

Braised Fish Heads with
Ginger and Scallions

薑蔥焗魚頭

薑蔥焗魚頭是非常下酒的順德菜，因所用的魚頭都是淡水魚，
所以要用薑、蔥、生抽、胡椒粉去除土味。

食材：
大魚頭或草魚頭700公克
薑片200公克
蔥段200公克

調味料：
生抽2茶匙
白糖2茶匙
精鹽1/2茶匙
胡椒粉1/2茶匙
麵粉200公克

作法：
1. 大魚頭或草魚頭剖開成8塊，備用。
2. 薑片、蔥段用熱油爆香，備用。
3. 起油鍋，將魚頭沾上麵粉，炸至金黃酥脆後撈起。
4. 另起油鍋燒熱，加入炸好的魚頭、薑片、蔥段、調味料炒勻後即可。

Braised Fish Heads with Ginger and Scallions

This is a delicacy of Shunde, a district of Foshan in Guangdong, China, where it has long developed a refined dining culture. This dish goes really well with liquors. It is made with freshwater fish, so ginger, scallions, light soy sauce and white pepper powder are used to deodorize the fishy smell.

Ingredients:
700g grass carp or other big fish heads
200g sliced ginger
200g sectioned scallions

Seasonings:
2 tsp light soy sauce
2 tsp sugar
1/2 tsp salt
1/2 tsp pepper powder
200g flour

Directions:
1. Cut the fish heads into 8 pieces and set aside.
2. Sauté the ginger and scallions until aromatic and set aside.
3. Heat oil in a wok. Coat the fish head pieces with flour and fry until golden brown and crisp. Remove and set aside.
4. Heat a bit of oil in another wok. Add the deep-fried fish head pieces, ginger, scallions and seasonings. Mix well and serve.

Yellow Croaker Jelly

黃魚凍

從前在菜市場很容易買到野生的小黃魚。我們家的阿嫂都會一大早去傳統市場挑
最新鮮的小黃魚，買個20條左右，回家煮好放涼，就會結凍。
這是我跟父親最喜歡吃的，做起來方便吃起來又鮮美，每一次我都會吃個3或4條。

食材：
小黃魚10條（每條約150公克～
200公克），或用豆仔魚也可
花椒10公克
八角10公克
薑片30公克
蔥段30公克

調味料：
高粱酒100cc
生抽100cc
老抽100cc
魚露100cc
白糖200公克
清水500cc

作法：
1. 小黃魚打鱗，將魚肚內的黑膜清乾淨。
2. 用薑片起鍋，快煎兩面。
3. 加入其餘食材及調味料，慢火煮20分鐘。
4. 待冷即成黃魚凍。

Yellow Croaker Jelly

It used to be very easy to find wild yellow croakers at farmers' markets. I remember the old
maid used to go to a traditional market in early morning to find the freshest yellow croakers
and come home with around 20 fish. She then cooked and waited for them to cool down.
My father and I loved the dish best. It was simple and savory. I could eat three or four
croakers at one time.

Ingredients:
10 yellow croakers or mullets
    (150-200g each)
10g Sichuan pepper
10g star anise
30g sliced ginger
30g sectioned scallions

Seasonings:
100ml Kaoliang spirit
100ml light soy sauce
100ml dark soy sauce
100ml fish sauce
200g sugar
500ml water

Directions:
1. Scale the croakers and scrape off the dark tissues in their bellies.
2. Sauté ginger in a wok. Add fish and quickly pan-fry on both sides of each fish.
3. Add the remaining ingredients and seasonings and cook at a lower temperature for 20 minutes.
4. Wait for it to cool down.

生抽煎大蝦碌

生抽煎大蝦是非常有名的海鮮，也有稱之為豉油皇大蝦。
在香港和大陸的醬油有分生抽及老抽，生抽鮮香，老抽濃香。
煎好的大蝦清脆鮮甜，加上少許的生抽，可增加蝦的甜度。
這是一道令人食指大動的請客大菜，但不能煎過熟，九分半是最好的。

食材：
大明蝦6隻
（約600公克左右）

調味料：
生抽3茶匙
白糖2茶匙

作法：
1. 大明蝦剪去腳鬚，開肚去腸泥，用布吸乾。
2. 起油鍋，放入少許油，將大明蝦慢煎至兩面呈金黃色。
3. 加入生抽、白糖翻炒，加蓋悶煮1分鐘即可。

Pan-fried Prawns in Light Soy Sauce

This is a very famous seafood delicacy in Hong Kong. There are two types of soy sauce—light and dark. The light one tastes thinner for its bean freshness. The dark one, however, is richer in flavor. The sautéed prawns is crisp in texture and delicious in flavor. A bit of light soy sauce simply enriches the taste. This is a great dish to entertain your guests, but beware not to overcook!

Ingredients:
6 prawns (about 600g)

Seasonings:
3 tsp light soy sauce
2 tsp sugar

Directions:
1. Trim off the prawn legs. Cut down the back of each prawn and remove intestinal tract. Dry them up with a towel.
2. Heat a bit of oil in a wok and fry the prawns at lower temperature until both sides turn golden brown.
3. Add light soy sauce and sugar. Stir-fry a bit and put a lid on the wok and continue to cook for 1 more minute.

Pan-fried Prawns in Light Soy Sauce

梅菜蔥段蒸鯇魚腩

客家人做的梅乾菜心非常香脆，有分鹹梅菜心及甜梅菜心。
如果你在香港或大陸買，記得要買甜梅菜心。
梅菜蔥段蒸鯇魚腩是非常道地的客家家常菜，食材簡單，作法快速，可口下飯。

食材：
草魚肚600公克
香港梅菜心150公克
薑片50公克
蔥段50公克

調味料：
生抽1茶匙
蠔油1茶匙
白糖1茶匙
沙拉油2湯匙

作法：
1. 草魚肚打鱗洗淨，清除魚肚內的黑膜，備用。
2. 梅乾菜洗淨切碎，入鍋炒香。
3. 加入薑片、蔥段及調味料拌好，放在魚背上，大火蒸12分鐘。
4. 端出加入少許熟油即可。

Steamed Grass Carp Belly with Pickled Mustard Greens and Sectioned Scallions

The Hakka people make nice and crisp pickled mustard greens. There are salty and sweet pickled mustards greens. Get the sweet ones when you purchase them in Hong Kong or in the Mainland. This is a truly genuine Hakka cuisine. The dish is simple and quick to make. It is tasty and goes well with steamed rice.

Ingredients:
600g grass Carp belly
150g pickled mustard greens
50g sliced ginger
50g sectioned scallions

Seasonings:
1 tsp soy sauce
1 tsp oyster sauce
1 tsp sugar
2 tbsp vegetable oil

Directions:
1. Scale and clean the fish belly and scrape off the dark tissue lining the abdominal cavity. Set aside.
2. Clean and mince the mustards greens. Toss them in a wok until fragrant.
3. Mix the mustard greens with ginger, scallions and seasonings. Place the mixture over the top of the fish and steam at a high temperature for 12 minutes.
4. Serve with a bit of boiled oil.

Steamed Grass Carp Belly with Pickled
Mustard Greens and Sectioned Scallions

陳皮薑米焗鯇魚腸

順德以河鮮出名。這道焗魚腸是非常香濃可口的菜餚，
但做起來有點繁複，魚腸一定要清洗乾淨，用鹽跟太白粉清洗最好。

食材：
草魚腸約800公克
陳皮碎10公克
薑米10公克
雞蛋3顆
清水2顆蛋的量

調味料：
生抽1/2茶匙
料理酒3茶匙
白糖1茶匙
精鹽1/2茶匙
白胡椒粉 1/2茶匙

作法：
1. 草魚腸用小剪刀剪開，用鹽洗淨，切成3公分長，吸乾水分。
2. 雞蛋3顆打散，加入清水、所有食材、調味料，攪拌均勻。
3. 倒入深盆中，蒸約15分鐘。要透氣，蛋才不會起蜂巢。

Steamed Grass Carp Intestines with Dried Tangerine Peels and Ginger Rice

Shunde is famous for its freshwater delicacies. This dish is very delicious but a bit complicated to make. The fish intestines must be cleaned thoroughly. It is better to do it with salt and potato starch.

Ingredients:
800g grass carp intestines
10g minced dried tangerine peels
3 eggs
Water (equivalent to 2 eggs)

Seasonings:
1/2 tsp light soy sauce
3 tsp cooking wine
1 tsp sugar
1/2 tsp salt
1/2 tsp pepper powder

Directions:
1. Cut open the intestines with a pair of small scissors. Clean them with salt and cut into 1-inch-long pieces. Dry up with a towel.
2. Beat the eggs. Mix the beaten eggs with water and the other ingredients and seasonings.
3. Pour the mixture into a deep bowl and steam for around 15 minutes. Remove the cover occasionally to avoid honeycomb-like texture.

Steamed Grass Carp Intestines with
Dried Tangerine Peels and Ginger Rice

陳皮豉汁蒸白鱔

鰻魚廣東人稱之為白鱔，白鱔是非常有活力的河鮮，有生機補充體力之功效。
處理白鱔要用熱水燙過，把皮上的滑液去除，後切厚片去燉。鰻魚用燉的能發揮很好的功效。

食材：
鰻魚1條600公克
老陳皮20公克
豆豉50公克
蒜茸50公克
蔥花10公克

調味料：
生抽1茶匙
蠔油1湯匙
白糖1茶匙
太白粉少許
精鹽少許

作法：
1. 鰻魚切成金錢狀，約1公分厚，備用。
2. 蒜茸炸至金黃色備用。
3. 陳皮、豆豉切碎炒香，加入蒜茸、調味料拌勻。
4. 加入鰻魚，大火蒸10分鐘後，撒上蔥花，淋上熟油即可。

Steamed Eel with Dried Tangerine Peels and Black Bean Sauce

Eels are called Baishan in Cantonese. They are vigorous river fish. They are believed to be able to bring life and energy to human bodies. Water blanching is required in order to get slime off eels that are then cut into thick pieces and cooked. Eel stews are believed to be good for health.

Ingredients:
1 eel (about 600g)
20g old dried tangerine peels
50g fermented black beans
50g finely minced garlic
10g chopped scallions

Seasonings:
1 tsp light soy sauce
1 tbsp oyster sauce
1 tsp sugar
A bit potato starch
A bit salt

Directions:
1. Cut the blanched eels into half-inch thick pieces and set aside.
2. Fry the minced garlic into golden brown.
3. Mince the tangerine peels and fermented black beans and stir-fry until fragrant. Add fried garlic and the seasonings and mix well.
4. Add eel pieces and steam at a high temperature for 10 minutes. Add scallions and pour boiled oil over the eel pieces.

薑蔥焗花蟹

這是一道鮮味方便的海鮮，每一家粵菜餐廳的廚師都會做。
煮的時候，也可加點廣東的伊麵，將生麵一起燜煮，讓麵吸收湯汁，吃起來非常可口。
如果你喜歡吃螃蟹，不妨照著我的食譜做來嚐嚐，蟹鮮汁甜，保證滋味無窮。

食材：
活花蟹600公克（其他活蟹亦可）
薑片300公克
蔥段300公克

調味料：
高湯400cc
白糖1茶匙
精鹽1/2茶匙
牛油1茶匙

作法：
1. 花蟹洗淨後切塊，拍上少許太白粉，入油鍋炸至金黃八分熟後撈起。
2. 用少許油起油鍋，放入薑蔥炒香。
3. 加入牛油、高湯、花蟹後加蓋燜煮15分鐘即可。

Braised Spotted Crabs with Ginger and Scallions

This is a simple and savory seafood delicacy that you can find in any Hong Kong restaurants. I-mien, a variety of flat Cantonese egg noodles made from wheat flour, can be added to this dish. The noodles, after absorbing the broth, taste very delicious. If you like crabs, you might as well try this recipe. The fresh crabs and luscious broth are a savory guarantee.

Ingredients:
600g live spotted crabs or other live crabs
300g sliced ginger
300g sectioned scallions

Seasonings:
400ml soup stock
1 tsp sugar
1/2 tsp salt
1 tsp butter

Directions:
1. Clean the crabs. Dust them with a bit of potato starch. Deep-fry in boiled oil until golden brown or until medium-well done.
2. Heat a little oil in a wok. Add ginger and scallions and toss until fragrant.
3. Add butter, soup stock and fried crabs. Put a lid on the wok and cook for 15 minutes.

薑蔥焗生蠔

這是一道非常營養、適合下飯下酒的好菜，一定要趁熱吃，否則生蠔冷了會有腥味。

食材：
冷凍進口生蠔600公克
薑片300公克
蔥段300公克

調味料：
生抽1茶匙
蠔油2茶匙
白糖1茶匙
鹽1/2茶匙
花雕酒2湯匙

作法：
1. 生蠔用太白粉洗淨備用。
2. 起油鍋，把薑蔥爆香。
3. 加入調味料、生蠔，加蓋燜煮10分鐘。
4. 開蓋加入少許太白粉水勾芡即可。

Braised Oysters with Ginger and Scallions

This dish has great nutrition. Serve hot with liquors or rice, otherwise cold oysters may taste fishy.

Ingredients:
600g imported frozen oysters
300g sliced ginger
300g sectioned scallions

Seasonings:
1 tsp light soy sauce
2 tsp oyster sauce
1 tsp sugar
1/2 tsp salt
2 tbsp Huadiao Jiu (a traditional Chinese alcoholic beverage)

Directions:
1. Mix the oysters with potato starch. Rinse well and set aside.
2. Heat oil in a wok, and sauté the ginger and scallions until aromatic.
3. Add the seasonings and oysters. Put a lid on the wok and cook for 10 minutes.
4. Remove the lid. Add potato starch water to thicken the broth.

Braised Oysters with Ginger and Scallions

豉汁炒蜆

這是一道平民美食。蜆是香港人的叫法,在台灣叫海瓜子,用海蛤蜊也可以。
香港的炒法跟台灣不太一樣,各有千秋。我年輕時在廟街最喜歡點這道菜,
一次可以吃五到六盤。雖說是家常菜,但味美鮮香,可稱得上是一道極品。

食材:　　　　　**調味料:**
蜆600公克　　　　沙拉油3湯匙
豆豉20公克　　　　生抽1湯匙
蒜茸20公克　　　　蠔油1湯匙
紅辣椒20公克　　　砂糖1湯匙

作法:
1. 蜆吐沙約3小時。
2. 起油鍋,放入豆豉、蒜茸、紅辣椒炒香。
3. 加入蜆、生抽、蠔油、砂糖翻炒約3分鐘,見蜆開殼便可食用。

Stir-Fried Clams in Black Bean Sauce

This is a popular dish in Hong Kong. People in Taiwan, however, call clams haiguazi or
haigeli. In Hong Kong, they are cooked in a way that is slightly different from the
Taiwanese-style stir-fried clams. In my memory as a young man, this was my favorite dish at
the Temple Street Night Market. I could devour as many as five or six dishes of clams in
Black Bean Sauce every time I went there. As popular as it may be, the sensory taste of the
dish stands for the finest.

Ingredients:　　　　　**Seasonings:**
600g clams　　　　　　　3 tbsp vegetable oil
20g fermented black beans　1 tbsp light soy sauce
20g finely minced garlic　　1 tbsp oyster sauce
20g red chili peppers　　　1 tbsp sugar

Directions:
1. Soak the clams in water for around 3 hours to spit out sand.
2. Heat oil in a wok. Add fermented black beans, garlic and red chili peppers. Stir-fry until fragrant.
3. Add clams, light soy sauce, oyster sauce and sugar and stir-fry for around three minutes or until shells pop open.

Stir-Fried Clams in Black Bean Sauce

鹹魚蓉鑲豆腐

鹹魚蓉鑲豆腐與鑲苦瓜同是客家菜的經典。帶有湯汁的炆豆腐，
拌在熱騰騰的白飯上，果真是吃得話都講不出來。

食材：
板豆腐600公克
梅花豬絞肉200公克
五花肉100公克
鹹魚蓉50公克

調味料：
生抽1茶匙
蠔油2茶匙
白糖1茶匙
太白粉1茶匙
高湯200cc

作法：
1. 板豆腐切成骨排大小，挖掉少許豆腐心，備用。
2. 鹹魚蓉用少許油炒香，加入豬肉、調味料攪拌好，鑲在豆腐上。
3. 將鑲好的豆腐煎至金黃色，不要煎太老嘍。加入高湯燜10分鐘，勾芡即可。

Tofu stuffed with Minced Salted Fish Paste

Like stuffed bitter gourds, this stuffed tofu delicacy is another classic Hakka cuisine. You can be too busy to say a word when you enjoy the juicy braised tofu and hot steamed rice in your mouth.

Ingredients:
600g firm tofu
200g ground pork collar-butt
100g pork belly
50g minced salted fish

Seasonings:
1 tsp light soy sauce
2 tsp oyster sauce
1 tsp sugar
1 tsp potato starch
200ml soup stock

Directions:
1. Slice the tofu into domino-size pieces. Slightly make a dent in the center of the tofu slices and then set aside.
2. Fry the salted fish with a bit of oil until fragrant. Mix it with pork and seasonings and stuff the tofu slices with the fish paste.
3. Fry the stuffed tofu until golden brown. Do not overcook. Add soup stock and braise for 10 minutes. Add potato starch water to thicken the soup.

*Steamed Minced Pork Patty
with Salted Fish*

鹹魚蒸肉餅

香港人都很喜歡吃鹹魚，尤其是我。
在香港要買好的生曬鹹魚，價錢很貴，通常我都會買黃花、馬友（即午魚），還有黑鯧魚。
鹹魚買回來不能洗，先切好你要用的大小，用玻璃罐裝好，將鹽放入瓶內，加滿沙拉油，泡約1星期就可用了。鹹魚蒸肉餅是很下飯的菜，我很喜歡。

食材：
鹹魚100公克（曹白或馬友鹹魚）
梅花豬肉400公克
薑絲50公克

調味料：
生抽1茶匙
蠔油1茶匙
白糖1/2茶匙
太白粉1茶匙

作法：
1. 梅花絞肉與所有調味料攪拌均勻成肉餅狀。
2. 將肉餅放到盤上，加入煎過之鹹魚在肉上。
3. 加少許白糖在鹹魚上，與薑絲蒸20分鐘即可。

Steamed Minced Pork Patty with Salted Fish

Many people in Hong Kong love salted fish. I especially love it. Good dried salted fish are very expensive in Hong Kong. The fish that I usually buy are yellow croakers, king salmons and black pomfrets. Do not rinse the salted fish. Cut them into smaller pieces and fill into a glass bottle. Add salt, and fill in vegetable oil. Wait for around one week, and it is done. This dish goes really well with steamed rice. I like it very much.

Ingredients:
100g salted fish
400g pork collar-butt
50g shredded ginger

Seasonings:
1 tsp light soy sauce
1 tsp oyster sauce
1/2 tsp sugar
1 tsp potato starch

Directions:
1. Mix ground pork with all the seasonings and form a meat patty.
2. Place the meat patty on a plate and add fried salted fish over the top of the meat patty.
3. Add a bit of sugar and shredded ginger on the salted fish and steam for 20 minutes.

鮮蠔汁鮑魚

食材：
鮮鮑魚12粒（約1200公克）
薑100公克
蔥100公克

調味料：
蠔油100公克
火腿汁200公克
白砂糖1湯匙
濃上湯500公克

作法：
1. 鮮鮑魚用開水燙熟，去殼，去腸。放入沙鍋，薑蔥調味料一同放入，以小火燜達1小時。
2. 收汁即可。

Abalones in Oyster Sauce

Ingredients:
12 fresh abalones (around 1,200g)
100g ginger
100g scallions

Seasonings:
100ml oyster sauce
200ml ham sauce
1 tbsp sugar
500ml thick stock

Directions:
1. Blanch the abalones in boiling water and remove their shells and intestines. Put the abalones, ginger, scallions and the seasonings in a casserole. Cook over low heat for 1 hour.
2. Cook until the broth thickens.

Abalones in Oyster Sauce

three

蔬菜和蛋

Vegetables and Eggs

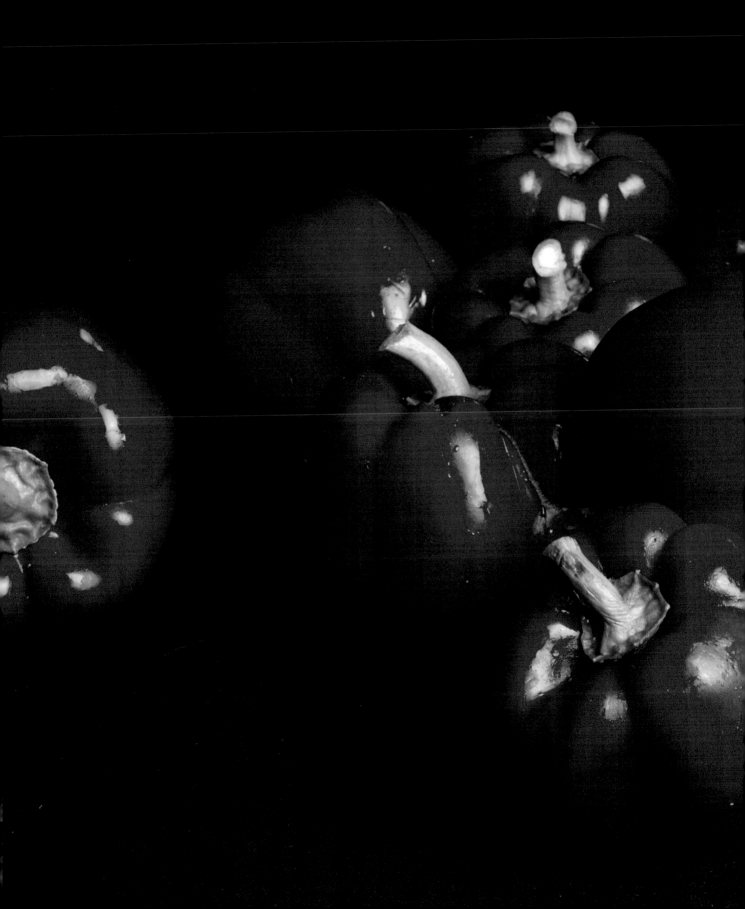

青瓜馬蹄炒圍蝦

做這道菜一定要用大黃瓜，大黃瓜炒至軟滑，再下圍蝦，清鮮甜。

食材：
大黃瓜一條約600公克
馬蹄200公克
沙蝦300公克
薑片30公克

調味料：
白糖1/2茶匙
精鹽1/2茶匙
太白粉1/2茶匙
高湯200cc

作法：
1. 大黃瓜去皮開邊去籽切厚片，馬蹄去皮切片。
2. 用少許油起油鍋，加入薑片、馬蹄、大黃瓜片、圍蝦炒香。
3. 放入高湯，燜焗約5分鐘。
4. 加入調味料，勾薄芡即可。

Stir-fried Cucumbers and Water Chestnuts with Shrimps

A large cucumber is required to make this delicacy. The cucumber is stir-fried until soft, and its flavor is enhanced by shrimps.

Ingredients:
1 large cucumber (about 600g)
200g water chestnuts
300g shrimps
30g sliced ginger

Seasonings:
1/2 tsp sugar
1/2 tsp salt
1/2 tsp potato starch
200ml soup stock

Directions:
1. Peel the cucumber and cut it open. Remove the seeds and cut into thick pieces. Skin water chestnuts and then slice.
2. Heat a bit of oil in a wok. Add ginger, sliced chestnuts, cucumber pieces and shrimps and toss until fragrant.
3. Add soup stock. Put a lid on the wok and cook at a low temperature for 5 minutes.
4. Add seasonings. Finally add a bit of potato starch water to thicken the broth.

蝦蜆肉炒韭菜花

這道菜是香港大排檔的小炒，講求鑊氣。大火快炒，色香味俱全。

食材：
沙蝦肉300公克
蜆肉300公克
韭菜花300公克
薑片20公克

調味料：
生抽1/2茶匙
白糖1/2茶匙
精鹽1/2茶匙
米酒1茶匙

作法：
1. 蝦肉、蜆肉洗淨備用，韭菜花洗淨切段。
2. 炒鍋燒熱，加入少許生油，放入薑片起鍋。
3. 隨即放入蝦肉爆香。
4. 再放入蜆肉、韭菜花炒香，加入調味料便可。

Stir-fried Chive Flowers with Shrimps and Clams

Wok heat matters when it comes to cook this daipaidong, or food stall, dish. Quickly tossed over high heat, the food is great in color, smell and taste.

Ingredients:
300g shelled shrimps
300g clam meat
300g chive flowers
20g sliced ginger

Seasonings:
1/2 tsp soy sauce
1/2 tsp sugar
1/2 tsp salt
1 tsp rice wine

Directions:
1. Clean shrimps and clams and section chive flowers.
2. Heat a bit of oil in a wok. Add sliced ginger.
3. Add shrimps and toss until aromatic.
4. Add clam meat and chive flowers and continue to toss until fragrant. Add seasonings and serve.

Stir-fried Chive Flowers with Shrimps and Clams

勝瓜雲耳炒洋蔥

角瓜清甜，雲耳爽脆，加上洋蔥的香味，這三合一是非常健康的絕配。

食材：
角瓜1條（約700公克）
洋蔥1個（約300公克）
雲耳30公克
薑片20公克
蔥段20公克

調味料：
生抽1茶匙
白糖1/2茶匙
精鹽1/2茶匙

作法：
1. 角瓜去皮洗淨切菱形，洋蔥去皮切頭尾一開八，雲耳泡水，備用。
2. 起油鍋，將油燒熱，炒香洋蔥。
3. 放入其餘食材炒2分鐘，加入調味料炒勻即可。

Zucchinis, black Mushrooms and Onions Stir Fry

The refreshing taste of zucchinis, the crispy texture of black mushrooms and the aromatic scent of onions together are a healthy fit.

Ingredients:
1 large zucchini (about 700g)
1 onion (about 300g)
30g black mushrooms
20g sliced ginger
20 sectioned scallions

Seasonings:
1 tsp Light soy sauce
1/2 tsp sugar
1/2 tsp salt

Directions:
1. Peel the zucchini and rinse. Cut it in half and then slice. Cut off the two ends of the onion and cut it into 8 pieces. Soak the black mushrooms and set aside.
2. Heat oil in a wok and stir-fry the onion pieces until aromatic.
3. Add the remaining ingredients and continue to toss for 2 more minutes. Add the seasonings and mix thoroughly.

蠔油涼瓜炒牛肉

涼瓜炒牛肉是夏天味的熱炒。涼瓜甘苦，牛肉香嫩，蠔油鮮，配白飯最佳。

食材：
牛柳肉400公克
苦瓜400公克
薑片10公克
蔥段10公克

調味料：
生抽1茶匙
蠔油2茶匙
白糖1茶匙
精鹽1/2茶匙
太白粉1茶匙

作法：
1. 牛肉切片，加入少許清水，用手打至牛肉吸夠水。
2. 苦瓜去籽去白色內膜，片薄片，以熱水汆燙後泡冰水。
3. 將牛肉用熱油炒至七分熟備用。
4. 苦瓜下鍋炒軟，加入調味料、牛肉翻攪，勾芡即可。

Stir-fried Beef Fillets and Bitter Gourds in Oyster Sauce

This is a hot stir-fry for summer. The bitter-sweet taste of bitter gourds, the tender texture of the beef and the umami flavor of the oyster sauce together create a dish that is perfect to go with steamed rice.

Ingredients:
400g beef fillets
400g bitter gourds
10 sliced ginger
10g sectioned scallions

Seasonings:
1 tsp light soy sauce
2 tsp oyster sauce
1 tsp sugar
1/2 tsp salt
1 tsp potato starch

Directions:
1. Slice beef and add a bit of water. Use hands to whip the beef until water absorbed.
2. Remove all the seeds and white tissue inside the bitter gourds. Finely slice the gourds. Blanch the sliced gourds and then cool in ice water.
3. Fry the beef in boiling oil until medium well. Set aside.
4. Stir-fry the sliced gourds until soft. Add seasonings and beef and continue to toss. Add potato starch water to give the final touch to the dish.

Bitter Gourds with Salted Fish Stuffing

鹹魚蓉釀苦瓜煲

客家菜裏有一道鹹魚蓉豬肉釀苦瓜，是正宗客家私房菜。
這道菜以我姊夫做的最好吃，因為他在香港開一家客家餐廳，叫醉瓊樓。

食材：
綠色大苦瓜2條（約800公克）
豬絞肉400公克
鹹魚蓉（馬友魚）30公克

調味料：
高湯500cc
生抽1茶匙
蠔油1湯匙
白糖1茶匙
精鹽1/2茶匙
太白粉2茶匙

作法：
1. 苦瓜切成2指寬圓形，將圓形內白膜切掉備用。
2. 豬絞肉拌上鹹魚蓉，加入生抽、白糖、太白粉攪拌，鑲入苦瓜內。
3. 起油鍋，放入苦瓜煎至兩面金黃色。
4. 加入調味料、高湯，小火悶煮20分鐘即可。

Bitter Gourds with Salted Fish Stuffing

This is a genuine Hakka delicacy. My brother-in-law is especially good at making this dish.
He is the owner of Tsui King Lau, a Hakka restaurant in Hong Kong.

Ingredients:
2 large green bitter gourds (about 800g)
400g ground pork
30g minced salted fish

Seasonings:
500ml Soup stock
1 tsp light soy sauce
1 tbsp oyster sauce
1/2 tsp sugar
2 tsp salt
A bit potato starch

Directions:
1. Cut the bitter gourds into 1-inch pieces of round shape. Remove the seeds and white tissue. Set aside.
2. Mix the ground pork with the salted fish. Add light soy sauce, sugar and potato starch and continue to mix evenly. Stuff the gourd pieces with the mixture.
3. Heat oil in a wok. Sauté the stuffed gourd pieces until both sides turn golden brown.
4. Add seasonings and soup stock. Put a lid on the wok and cook at a lower temperature for 20 minutes.

薑汁鹹魚蓉炒芥蘭

遇到胃口不佳的時候，建議你做一道鹹魚炒芥蘭，非常下飯。

食材：
芥蘭菜1.5公斤
鹹魚蓉30公克

調味料：
高粱酒2茶匙
薑汁1茶匙
鹽1/2茶匙

作法：
1. 將芥蘭菜洗淨，只取菜葉部位。
2. 鹹魚蓉半煎炸香備用。
3. 用少許沙拉油起油鍋，油溫達100℃時下芥蘭菜翻炒。
4. 加入高粱酒、薑汁、鹹魚蓉，加蓋燜2分鐘，翻炒即可上盤。

Stir-fried Kai-lan with Ginger Juice and Minced Salted Fish

When you have a poor appetite, you might as well try this dish that makes you eat more rice.

Ingredients:
1.5kg Kai-lan, aka Chinese broccoli
50g minced salted fish

Seasonings:
2 tsp Kaoliang spirit
1 tsp ginger juice
1/2 tsp salt

Directions:
1. Rinse the Chinese broccoli. Select only the tender parts.
2. Fried the salted fish until fragrant. Set aside.
3. Heat a bit of vegetable oil in a wok to 100℃. Stir-fry the Chinese broccoli.
4. Add Kaoliang spirit, ginger juice and salted fish. Put a lid on the wok and cook for 2 more minutes. Stir-fry a bit and serve.

Stir-fried Kai-lan with Ginger Juice and Minced Salted Fish

蘿蔔煮魚餅

冬天的蘿蔔特別鮮甜，加上帶點湯汁的魚餅，味道鮮美，是奶媽常煮給我吃的一道名菜。

食材：
魚漿300公克
碎臘腸50公克
白蘿蔔500公克
蔥花50公克
薑片20公克

調味料：
胡椒粉1/2茶匙
生抽1/2茶匙
蠔油1茶匙
白糖1/2茶匙
太白粉2茶匙
精鹽1/2茶匙

作法：
1. 魚漿加入精鹽、太白粉、臘腸，攪拌好備用。
2. 白蘿蔔切呈長條狀，約食指大小，備用。
3. 起油鍋，把魚漿煎成金黃色的魚餅後取出，切成條狀食指大小。
4. 將蘿蔔炒香、放入蔥、調味料，加入500cc水，煮10分鐘。
5. 放入魚餅，加蓋煮5分鐘。
6. 用太白粉勾薄芡後即可。

Fish Cake and Radish

Radishes taste best in winter, especially when they are cooked with fish cakes in a broth.
My nanny used to cook this famous dish for me when I was little.

Ingredients:
300g fish paste
50g minced sausages
500g Chinese radishes
50g chopped scallion
20g sliced ginger

Seasonings:
1/2 tsp pepper powder
1/2 tsp light soy sauce
1 tsp oyster sauce
1/2 tsp sugar
2 tsp potato starch
1/2 tsp salt

Directions:
1. Add the salt, potato starch and sausages into the fish paste. Mix them well and set aside.
2. Cut the radishes into 3-inch strips and set aside.
3. Heat oil in a wok. Pan-fry the fish paste until it turns golden brown. Cut it into 3-inch strips.
4. Fry the radish strips until fragrant. Add scallions, seasonings and 500ml of water and cook for 10 minutes.
5. Add the stripped fish cake. Put a lid on the wok and cook for another 5 minutes.
6. Add a bit of potato starch water to thicken the broth.

Fish Cake and Radish

蜆肉菜脯煎蛋

我相信大家都吃很多菜脯蛋，但是你不見得吃過加入生開海瓜子肉一道去煎的，味道鮮到讓你吃了眼淚都會流出來呀！

食材：
蜆肉300公克
碎菜脯200公克
雞蛋3顆
蔥花50公克

調味料：
生抽少許
白糖少許
精鹽少許
太白粉10公克

作法：
1. 碎菜脯洗淨，起鍋炒香備用。
2. 雞蛋打散，加入菜脯、蜆肉、蔥花調味料拌勻。
3. 起油鍋，將蛋液倒入，煎至兩面呈現金黃色即可。

Omelet with Dried Radishes and Clams

Dried Radish Omelet may not be an uncommon dish in Taiwan; however, it can be a delightful surprise to your taste buds when clams are added. The dish is so delicious that it may even bring tears of joy to your eyes!

Ingredients:
300g clam meat
200g minced dried radishes
3 eggs
50g chopped scallion

Seasonings:
A bit Light soy sauce
A bit sugar
A bit salt
10g potato starch

Directions:
1. Rinse the dried radishes. Heat a wok and add the radishes. Toss until fragrant.
2. Beat the eggs and add the dried radishes, clams and seasonings. Mix well.
3. Heat oil in a wok. Pour in the egg mixture and fry until golden brown on both sides.

Steamed Eggs with Crabs

奄仔蟹蒸水蛋

香港人所謂的奄仔蟹是指未經交配的螃蟹，也就是台灣人說的處女蟳。
夏秋之交正是處女蟳是最肥的時候，處女蟳蒸水蛋滑嫩鮮美，
因蟹汁以融入蒸蛋中，用來拌飯是一流。

食材：
小隻處女蟳（約600公克）
雞蛋3顆
清水3顆蛋的量
薑汁10公克

調味料：
生抽1茶匙
鹽1/3茶匙
麻油2滴

作法：
1. 處女蟳洗淨切開，蟹身切成6件，蟹鉗用刀背撬開蟹蓋備用。
2. 雞蛋中加入清水、薑汁、生抽、鹽、麻油打散。
3. 將蟹加放在上，慢火蒸15分鐘即可，每5分鐘要疏氣。

Steamed Eggs with Crabs

Unspawned female crabs are called chunuxun, or virgin crabs, in Taiwan. The transition time between summer and fall is when the crabs grow best. This dish tastes silky in texture and delicious in flavor. The crab broth enriches the taste of steamed eggs that goes really well with rice.

Ingredients:
Small virgin crabs (about 600g)
3 eggs
Water (equivalent to the eggs)
10g ginger juice

Seasonings:
1 tsp light soy sauce
1/3 tsp salt
2 drops sesame oil

Directions:
1. Rinse the crabs well and cut each of them into 6 parts. Crack the claws with the back of a knife blade and save the shell for later use.
2. Add water, ginger juice, light soy sauce, salt and sesame oil into the eggs and beat well.
3. Add the crabs in the egg mixture and steam at a lower temperature for 15 minutes. Remove the cover every 5 minutes to let out the steam.

番茄煮雞蛋

我從小就喜歡吃番茄煮雞蛋、牛肉或魚。
番茄的茄紅素對身體健康極好，加上雞蛋、牛肉或魚都富含蛋白質，
所以這道菜的營養價值非常高，花的錢又少，喜歡吃米飯的朋友不妨多煮這道菜。

食材：
番茄4顆（約400公克）
雞蛋4顆
薑片50公克
蔥段50公克

調味料：
番茄醬100公克
白糖2茶匙
鹽1茶匙

作法：
1. 雞蛋加入少許鹽打散，用少許油煎炒一下。
2. 待香味傳出，剛熟時用鍋鏟把蛋切開後盛起。
3. 番茄用熱水煮熟去皮，一開八。
4. 用薑片、油將番茄炒香炒軟。
5. 加入番茄醬、白糖、鹽、炒蛋，煮5分鐘即可。

Sauted Tomatoes and Eggs

I have always loved stir-fried tomatoes with eggs, beef or fish. Lycopene is very good for health. Eggs, beef or fish are rich in protein. This dish provides a high nutritious value at a lower cost. Those who love steamed rice might as well make this dish more often.

Ingredients:
4 tomatoes (about 400g)
4 eggs
50g sliced ginger
50g sectioned scallion

Seasonings:
100g ketchup
2 tsp sugar
1 tsp salt

Directions:
1. Add salt to the eggs and beat it well. Sauté the beaten eggs with a bit of oil.
2. Wait until fragrant and use spatula to break up the fried eggs.
3. Cook tomatoes in boiling water. Skin and cut each one into 8 pieces.
4. Add sliced ginger and oil in a wok. Add tomato and sauté until fragrant and soft.
5. Add ketchup, sugar, salt, fried eggs and cook for another 5 minutes.

鹹蛋皮蛋蒸肉餅水蛋

夏天到了，胃口不好，我就請奶媽做一份碎肉蒸三蛋。
是哪三蛋？就是鹹蛋、皮蛋和雞蛋打在一起，加上碎豬肉，用來拌飯清爽又滑口。

食材：
生鹹蛋2顆
生皮蛋1顆
生雞蛋3顆
豬絞肉100克

調味料：
生抽少許
白糖少許
麻油1/3茶匙

作法：
1. 皮蛋切粒，鹹蛋白取出備用。
2. 雞蛋與鹹蛋白先打發，再加入豬絞肉、100克水、調味料拌打均勻。
3. 加入皮蛋與鹹蛋黃，蒸約15分鐘，要疏氣，才不會起蜂巢。

Steamed Pork Patty with Salted, Preserved and Fresh Eggs

When I had a poor appetite in summer, my nanny used to make a steamed pork patty with three types of eggs for me. What kinds of eggs are they? They are salted, preserved and fresh eggs. Mix the eggs with ground pork. It is a delight to your taste buds when you have it with rice.

Ingredients:
2 salted eggs
1 preserved egg
3 eggs
100g ground pork

Seasonings:
A bit light soy sauce
A bit sugar
1/3 tsp sesame oil

Directions:
1. Dice the preserved egg. Remove the salted egg whites and set aside. Dice the salted yolks.
2. Beat the fresh eggs with salted egg white until frothy. Add ground pork, 100g water and seasonings and continue to beat evenly.
3. Add diced preserved egg and salted yolks and steam for 15 minutes. Remove the cover occasionally to avoid honeycomb-like texture.

four

煲湯和麵飯

Soup, Noodles and Rice

眉豆花生豬骨煲雞腳湯

眉豆與花生具健腦去濕的功效，加上豬骨、雞腳，營養好，膠質夠，是秋冬最適合的煲湯。

食材：
眉豆100公克
有衣花生150公克
豬骨300公克
雞腳400公克
老薑30公克

調味料：
鹽少許
清水300cc

作法：
1. 把所有食材洗淨備用。
2. 龍骨、雞腳先用開水去除血水後，放入大砂鍋。
3. 加入所有食材，慢火煲3小時，加鹽調味。

Black-eyed Peas, Peanuts, Pork Bones and Chicken Feet Stew

Purple haricot and peanuts are good for legs, brain and dehygrosis. Pork bones and chicken feet are nourishing and rich in colloid. Fall and winter are the best seasons to savor this clay-pot dish.

Ingredients:
100g black-eyed peas
150g peanuts with skin
300g pork bones
400g chicken feet
30g ginger

Seasonings:
A bit salt
300ml water

Directions:
1. Rinse all the ingredients and set aside.
2. Blanch pork back bones and chicken feet and put them in a large casserole.
3. Add all the other ingredients and cook at a lower temperature for 3 hours. Add salt and serve.

Black-eyed Peas, Peanuts,
Pork Bones and Chicken Feet Stew

花膠老雞煲响螺頭

花膠與响螺都是滋陰補腎，加上用老雞煲的湯，非常鮮甜可口。
這道湯富含膠質，養顏美容，不論男女老少都適合。

食材：
活响螺頭400公克
水發花膠200公克（後下）
老雞600公克
豬腱肉300公克
老薑30公克
清水4000cc

調味料：
鹽少許

作法：
1. 老雞、豬腱用開水煮10分鐘去血水。
2. 响螺肉先用清水慢火煲2小時，再放入老雞、豬腱肉、老薑煲1小時。
3. 加入花膠煲半小時，用鹽調味即可。

Stewed Chicken with Fish Maws and Conches

Fish maws and conches are nourishments to yin and yang energies that are present in our bodies. When they are cooked in chicken soup, they make the soup so hearty and flavorful. This soup is rich in colloid and good for skin. This is a dish for everyone.

Ingredients:
400g live conches
200g soaked fish maws
1 mature chicken (about 600g)
300g pork hock
30g ginger
4000ml water

Seasonings:
A bit salt

Directions:
1. Blanch the chicken and pork hock in boiling water and cook for 10 minutes.
2. Cook conches in water at a low temperature for 2 hours. Add blanched chicken, pork hock and ginger and continue to cook for another 1 hour.
3. Add flower maws and cook for another 0.5 hour. Add salt and serve.

青紅蘿蔔無花果煲牛坑腩

青蘿蔔與紅蘿蔔是香港人最喜歡的一道煲湯，方便快速。
這道湯甘甜味鮮，又有消腫止瀉利尿的功效，是四季皆宜的好湯。

食材：
青蘿蔔300公克
紅蘿蔔300公克
乾無花果10粒
牛坑腩800公克
老薑30公克
清水3000cc

調味料：
鹽少許

作法：
1. 牛坑腩用開水滾15分鐘去血水後，用冷水清洗，放入大砂鍋內。
2. 青紅蘿蔔切好，老薑、無花果洗淨，放入大砂鍋內。
3. 加入清水煲3小時，用鹽調味即可。

Stewed Beef Brisket with Green Radishes, Carrots and Figs

This is one of the most popular soups in Hong Kong because it is easy and quick to make. It is not only great in flavor but also good in medical functions such as reducing the swelling, stopping the runs and increasing urine production. This is a nourishing soup for all seasons.

Ingredients:
300g green radishes
300g carrots
10 dried figs
800g beef brisket
30g ginger
3000ml water

Seasonings:
A bit salt

Directions:
1. Blanch beef brisket in boiling water for 15 minutes and then rinse in cold water. Put it in a large casserole.
2. Cut green radishes and carrots into chunks. Rinse ginger and figs. Add all of them in the casserole.
3. Add water and cook at a lower temperature for 3 hours. Add salt and serve.

Stewed Beef Brisket with Green Radishes, Carrots and Figs

章魚蜜棗陳皮豬腱煲蓮藕

蓮藕有補心益腎滋陰養血的功效，加入乾的章魚，即八爪魚，
新鮮豬腱肉的口感，陳皮的香氣及化痰，是夏天的聖品。

食材：
乾八爪魚2隻
蜜棗10粒
陳皮5公克
蓮藕400公克
豬腱肉600公克
清水3000cc

調味料：
鹽少許

作法：
1. 豬腱肉用開水煮10分鐘去血水，用冷水沖涼，放入大砂鍋內。
2. 加入洗淨之食材，慢火煲3小時，用鹽調味即可。

Stewed Pork Hock and Lotus Roots with Octopuses, Candied Dates and Dried Tangerine Peels

Lotus roots are good for hearts and kidneys and are believed to be able to nourish yin energy and tonify the blood. Dried octopuses enrich the flavor. Pork hock provides springy texture. Dried tangerine peels offer fragrance and reduce sputum. This is a great dish for summer.

Ingredients:
2 dried octopuses
10 candied dates
5g dried tangerine peels
400g lotus roots
600g pork hock
3000ml water

Seasonings:
A bit salt

Directions:
1. Blanch the pork hock in boiling water and cook for 10 minutes. Rinse the blanched pork under running tape water and then put it in a large casserole.
2. Add the rinsed remaining ingredients in the casserole and cook at a lower temperature for 3 hours. Add salt and serve.

Stewed Pork Lungs with Sweet and
Bitter Almonds and Dried Vegetables

南北杏菜乾煲豬肺

煲豬肺補肺，有止咳化痰調肺的功效。這是一道非常清熱的湯品，但製作過程繁複。

食材：

南杏80公克	豬骨300公克
北杏30公克	豬腳300公克
菜乾200公克	雞腳300公克
陳皮5公克	清水3000cc
豬肺2個	

調味料：
鹽少許

作法：
1. 豬肺用水管插入喉管沖水放水數次，至完全洗淨，再用熱水燙10分鐘去血水及泡泡，擦乾水備用。
2. 豬骨、雞腳煮10分鐘去血水。
3. 將所有食材放入大砂鍋內，加入清水煲3小時，用鹽調味即可。

Stewed Pork Lungs with Sweet and Bitter Almonds and Dried Vegetables

Stewed lungs are good for lungs. They are believed to relieve coughs and reduce sputum. This clay-pot dish helps clear the heat in human bodies although the preparation can be complicated.

Ingredients:

80g sweet almonds	300g pork bones
30g bitter almonds	300g pork feet
200g dried vegetables	300g chicken feet
5g dried tangerine peels	3000ml water
2 pork lungs	

Seasonings:
A bit Salt

Directions:
1. Pipe water into the pork lungs several times until they are thoroughly clean. Blanch the pork lungs in boiling water and cook for 10 minutes. Dry them up with a towel.
2. Blanch pork bones and chicken feet in boiling water and cook for 10 minutes.
3. Put all the ingredients in a large casserole. Add water and cook for 3 hours. Add salt and serve.

八寶冬瓜湯

這是夏天散熱的一道燉湯。夏天食欲不佳，不妨來一碗清鮮的湯水保養。

食材：

冬瓜1/2個　　　　烤鴨80公克
蟹肉80公克　　　鴨腎80公克
冬菇粒80公克　　草菇80公克
水發瑤柱80公克　薑片
雞肉粒80公克　　高湯1200公克

調味料：
鹽少許

作法：
1. 把冬瓜洗淨，去籽去白膜。
2. 加入所有食材，蒸40分鐘，用鹽調味即可挖冬瓜肉吃。

Stewed Wax Gourd with "Eight Treasures"

This summer delicacy helps clear the heat in human bodies. When you have a poor appetite in summer, you might as well give yourself a bowl of freshness to nourish your body.

Ingredients:

1/2 wax gourd　　　　　80g diced roast duck
80g crab meat　　　　　80g diced duck kidneys
80g diced mushroom　　80g straw mushroom
80g soaked scallops　　 Sliced ginger
80g diced chicken meat　1200ml soup stock

Seasonings:
A bit salt

Directions:
1. Rinse wax gourd and remove internal seeds and white tissue.
2. Add all the ingredients and steam for 40 minutes. Add salt and serve.

Stewed Wax Gourd with
"Eight Treasures"

西洋菜陳腎煲龍骨

陳腎是廣東人的叫法，也叫臘鴨腎，與西洋菜、豬骨同煲（也可煲生魚），
清熱益肺，是適合夏天喝的好湯。

食材：
西洋菜600公克
豬龍骨600公克
陳皮10公克
乾鴨腎100公克
薑片30公克
清水3000cc

調味料：
鹽少許

作法：
1. 豬骨用開水煮10分鐘去血水，放入大砂鍋內。
2. 加入洗淨的所有食材和清水，慢火煲3小時，用鹽調味即可。

Stewed Pork Backbones and Roast Duck Kidneys with Watercress

Roast duck kidney is called Chan San in Cantonese. When it is cooked with watercress and pork backbones or fish, it is believed to clear heat and nourish lungs. It is a healthy soup for summer.

Ingredients:
600g watercress
600g pork backbones
10g dried tangerine peels
100g roast duck kidneys
30g sliced ginger
3000ml water

Seasonings:
A bit salt

Directions:
1. Blanch pork backbones in boiling water and cook for 10 minutes. Put the blanched pork bones in a large casserole.
2. Add all rinsed ingredients and water and cook at a lower temperature for 3 hours. Add salt and serve.

Stewed Pork Backbones and Roast
Duck Kidneys with Watercress

生木瓜陳皮老薑煎魚湯

青木瓜含維生素 A 及木瓜酵素，還可幫助乳腺發育。
這道湯加上煎過的鮮魚，很適合夏天配飯吃。
香港人認為，產後或開刀後喝這道湯最合適。
此外，它也有改善乾燥肌膚的效果。

食材：
生木瓜400公克
海鱸魚600公克
老薑50公克
陳皮5公克
清水3000cc

調味料：
鹽少許
胡椒粉少許

作法：
1. 鮮魚去鱗洗淨，用少許油煎至兩面金黃色備用。
2. 生木瓜去皮去籽，切成骨牌狀。
3. 清水燒開，放入所有食材，中火滾半小時至乳白色，加鹽調味即可。

Stewed Fish with Green Papaya, Dried Tangerine Peels and Ginger

Papaya is rich in Vitamin A and enzymes. It is widely believed as a breast enhancement fruit. This fried fish soup is good to go with rice in summer. It is good nourishment for postnatal women and for people who have just gone through surgeries. It is also a soothing relief to dry skin.

Ingredients:
400g Green papaya
600g sea basses
50g ginger
5g dried tangerine peels
3000ml water

Seasonings:
A bit salt
A bit pepper powder

Directions:
1. Scale the fish and rinse. Fry the fish with a bit of oil and cook until golden brown on both sides. Set aside.
2. Skin the papayas and then slice into domino-size pieces.
3. Put all the ingredients in boiling water and cook for half an hour or until the soup turns milky white. Add salt and serve.

Stewed Fish with Green Papaya,
Dried Tangerine Peels and Ginger

白胡椒老薑煲豬肚龍骨

自古以來，中國人就有一套以形補形的食補觀念，煲豬肚是補胃，
豬肚沾上醬油吃來十分可口。胡椒可暖胃，所以此湯應是適合秋冬的煲湯。

食材：
白胡椒粒30公克
老薑50公克
生豬肚2個
豬龍骨400公克
清水4000cc

調味料：
鹽少許

作法：
1. 豬骨洗淨去血水備用。
2. 豬肚翻開，用精鹽擦洗乾淨，用熱水煮20分鐘取出，與豬骨一同沖冷水。
3. 將豬骨與豬肚放入大砂鍋內，加入清水、老薑、胡椒粒，慢火煲4小時，加鹽調味即可。

Stewed Pork Stomachs and Backbones with White Pepper and Ginger

The Chinese people have long believed that you are what you eat. Stewed pork stomach is believed to be good for stomach. Pork stomachs are quite tasty with soy sauce. White pepper keeps stomach warm, so this soup is perfect for fall and winter!

Ingredients:
30g white peppercorns
50g ginger
2 pork stomachs
400g pork backbones
4000ml water

Seasonings:
A bit salt

Directions:
1. Clean the pork stomachs well and set aside.
2. Turn the stomachs inside out. Rub them with salt. Blanch pork stomachs and backbones in boiling water for 20 minutes and then rinse in cold water.
3. Put pork backbones and stomachs in a casserole. Add water, ginger and peppercorns and cook at a lower temperature for 4 hours. Add salt and serve.

赤小豆粉葛沙參玉竹煲魚湯

食材：

赤小豆100公克　　老薑50公克
沙參50公克　　　虱目魚2條
玉竹50公克　　　豬骨400公克
粉葛或山藥400公克　清水4000cc

調味料：

鹽少許

作法：

1. 先將魚去鱗去腸，洗淨，煎至金黃色，備用。
2. 將所有食材洗淨，放入大砂鍋。
3. 加入清水，慢火煲3小時，加鹽調味即可。

Fish Soup with Red Beans, Acorus Gramineus, Sea Cucumbers and Polygonatum Odoratum

Ingredients:

100g red beans
50g sea cucumbers
50g polygonatum odoratum
400g acorus gramineus or Chinese yam
50g ginger
2 milkfish
400g pork bones
4000ml water

Seasonings:

A bit salt

Directions:

1. Scale the fish and remove the entrails. Rinse well. Fry the fish until golden brown and set aside.
2. Rinse all the ingredients and put them in a large casserole.
3. Add water and cook at a lower temperature for 3 hours. Add salt and serve.

Fish Soup with Red Beans, Acorus Gramineus,
Sea Cucumbers and Polygonatum Odoratum

豉油皇炒麵

豉油皇炒麵是香港路邊經典的小吃，早、午、晚餐都可以吃。
如果早餐上粥店，可炒麵配白粥；吃午餐可做主食。
豉油的香氣，加上蛋麵的口感，真是經濟又實惠的一餐。

食材：
細蛋麵400公克
韭黃100公克
芽菜100公克
芝麻1茶匙
洋蔥絲100公克

調味料：
生抽2茶匙
老抽2茶匙（無鹽）
白糖1/2茶匙

作法：
1. 洋蔥切絲，韭黃切段，芽菜備用。
2. 將洋蔥炒熟備用，蛋麵炒香。
3. 加入調味料拌炒，再放入其他食材翻炒即可。

Fried Egg Noodles in Soy Bean Sauce

This is a classic street food in Hong Kong. It is a dish for every meal. You may order this dish with congee for breakfast at a rice porridge restaurant. It can also be a lunch entrée. The aroma of black bean sauce and the chewing texture of egg noodle together create an economic meal.

Ingredients:
400g thin egg noodles
100g hotbed chives
100g bean sprouts
1 tsp sesame seeds
100g shredded onions

Seasonings:
2 tsp light soy sauce
2 tsp dark soy sauce (unsalted)
1/2 tsp sugar

Directions:
1. Shred onions, section chives and rinse bean sprouts. Set aside.
2. Stir-fry onion well. Add noodles and continue to toss until fragrant.
3. Add seasonings and the rest of the ingredients. Toss well.

Fried Egg Noodles in Soy Bean Sauce

Crab Rice Porridge

水蟹粥

廣東潮汕人喜愛水蟹粥，因蟹煮粥鮮美可口，夏天沒胃口的時候吃一碗很清爽。

食材：
活海蟹2隻（約800公克）
白米約1碗
薑片50公克
蔥花20公克
大頭菜50公克

調味料：
鹽少許
胡椒粉少許

作法：
1. 活蟹殺好斬件備用。
2. 白米洗淨，6碗水煮至米開花。
3. 放入螃蟹、薑片，滾15分鐘。
4. 放入蔥花、切碎大頭菜、鹽、胡椒粉調味即可。

Crab Rice Porridge

The Chaoshan people from Guangdong province love this dish. The fresh crabs make the congee so palatable that it helps overcome the poor appetite in summer.

Ingredients:
2 live sea crabs (about 800g)
Rice (1 bowl or so)
50g sliced ginger
20g chopped scallions
50g turnips

Seasonings:
A bit salt
A bit pepper powder

Directions:
1. Chop the crabs into parts. Set aside.
2. Rinse rice and boil in 6 bowls of water until rice broken up into small pieces.
3. Add chopped crabs and sliced ginger and continue to cook for another 15 minutes.
4. Add chopped scallions, minced turnip, salt and pepper powder.

基圍蝦田雞粥

身體不適或精神不濟，如果來一碗粥，真是精神為之一振。

食材：
沙蝦500公克
田雞腿3只（約500公克）
薑片少許
蔥花少許
大頭菜絲少許
白米約1碗
水6碗

調味料：
鹽少許
胡椒粉少許
生抽少許
太白粉少許

作法：
1. 沙蝦剪去鬚腳。
2. 田雞去皮斬件，先洗淨，用少許油、太白粉拌一下。
3. 白米加水煮至米開花，放入薑片、沙蝦、田雞，滾5分鐘。
4. 撒上蔥花、大頭菜絲即可。

Shrimp and Frog Rice Porridge

When one feels fatigue, having a bowl of rice porridge always helps bring energy back.

Ingredients:
500g sand shrimps
3 frog legs (about 500g)
A bit sliced ginger
A bit chopped scallions
A bit shredded turnips
Rice (1 bowl or so)
Water (6 bowls)

Seasonings:
A bit salt
A bit pepper powder
A bit light soy sauce
A bit potato starch

Directions:
1. Trim shrimp legs.
2. Chop frog legs into parts and rinse. Marinate the legs in a bit oil and potato starch.
3. Cook rice in water until broken up. Add sliced ginger, shrimps and frog legs and continue to cook for another 5 minutes.
4. Add chopped scallions and shredded turnip.

Shrimp and Frog Rice Porridge

豬油雞蛋豉油撈飯

我小的時候常常吃飛機餐，所謂飛機餐就是快速的意思。
請奶媽為我燙一盤菜心式芥蘭，煮一碗很熱的白飯，放入自家炸的豬油，
打入新鮮的土雞蛋，淋上醬油一拌，真是天下無敵呀！

食材：
熟豬油1/2湯匙
生雞蛋1顆
生抽1/2湯匙
煮得很熱的白飯

作法：
1. 將熟豬油、生雞蛋、生抽放於大碗內。
2. 再將剛煮好的熱飯倒入碗中拌好即可。

Steamed Rice Mixed with Lard, Eggs and Soy Sauce

This is a "fast food" that I used to enjoy when I was little. I had my nanny make me a dish of scalded Chinese broccoli and a bowl of hot steamed rice mixed with home-made lard, fresh chicken eggs and soy sauce. Nothing can beat this gourmet taste!

Ingredients:
1/2 tbsp cooked lard
1 fresh egg
1/2 tbsp light soy sauce
Hot steamed rice

Directions:
1. Put the cooked lard, fresh egg and light soy sauce in a big bowl.
2. Add hot steamed rice and mix well.

Steamed Rice Mixed with Lard,
Eggs and Soy Sauce

· 右方為上海醉月樓主廚 戴忠麟

A Gourmet Chef's Home Cooking

親愛的讀者：
感謝您購買《憶猶未盡之住家菜》一書，希望您在看完這本書，或是開始嘗試
學做裏面的菜之後，填寫這張問卷調查表，並將此問卷調查表寄回。
您寶貴的意見，將是我們未來改進的動力。

1 您從何處購得本書?
□博客來網路書店 □金石堂網路書店 □誠品網路書店 □其他網路書店
□實體書店_____

2 您從何處得知本書?
□廣播媒體 □實體書店 □網路書店 □臉書 □朋友推薦
□實體書店_____

3 您購買本書的因素有哪些？（可複選）
□作者 □內容 □圖片 □版面編排 □其他_____

4 您覺得本書的封面設計如何？
□非常滿意 □滿意 □普通 □很差 □其他_____

5 非常感謝您購買此書，您還對哪些主題有興趣？（可複選）
□異國料理 □麵包烘焙類 □甜鹹點心類 □飲品類 □瘦身美容
□養生保健 □兩性關係 □心靈療癒 □其他_____

6 您最常選擇購書的通路是以下哪一個？
□誠品實體書店 □金石堂實體書店 □博客來網路書店 □誠品網路書店 □金石堂網路書店
□PC HOME網路書店 □Costco □其他網路書店_____ □其他實體書店_____

7 若本書出版形式為電子書，您的購買意願？
□會購買 □不一定會購買 □視價格考慮是否購買 □不會購買 □其他_____

8 您是否有閱讀電子書的習慣？
□有，已習慣看電子書 □偶爾會看 □沒有，不習慣看電子書
□其他_____

您認為本書尚需改進之處？以及對我們的意見？

9 _____
_____
_____
_____

10 日後若有優惠訊息，您希望我們以何種方式通知您？
□電話 □E-mail □簡訊 □書面宣傳寄送至貴府 □其他_____

謝謝您提供寶貴的意見，
您填妥寄回後，將我們將不定期提供
最新的會訊與優惠活動資訊給您：

姓名_____ 出生年月日_____
電話_____ E-mail_____
通訊地址_____

地址： 　　縣/市　　　鄉/鎮/市/區　　　路/街
　　　　段　　巷　　弄　　號　　樓

廣　告　回　函
台北郵局登記證
台北廣字第2780號

SAN YAU

三友圖書有限公司　收
SANYAU PUBLISHING CO., LTD.

106　　台北市安和路2段213號4樓

SAN YAU

三友圖書 / 讀者俱樂部

填妥本問卷，並寄回，即可成為三友圖書會員。
我們將優先提供相關優惠活動訊息給您。

優質好康

粉絲招募
歡迎加入

○ 看書 所有出版品應有盡有
○ 分享 與作者最直接的交談
○ 資訊 好書特惠馬上就知道

旗林文化╳橘子文化╳四塊玉文化
https://www.facebook.com/comehomelife

憶猶未盡之住家菜
A Gourmet Chef's Home Cooking

作　　者／劉冠麟

譯　　者／Li-Yu Chen

攝　　影／蕭維剛

發 行 人／程安琪

總 策 劃／程顯灝

編輯顧問／錢嘉琪

　　　　　潘秉新

總 編 輯／呂增娣

執行主編／李瓊絲

編　　輯／李雯倩

　　　　　吳孟蓉

　　　　　程郁庭

企劃行銷／謝儀方

美術總監／洪瑞伯

封面內文／洪瑞伯

出版者／橘子文化事業有限公司

總代理／三友圖書有限公司

地　址／106台北市安和路二段213號4樓

電　話／（02）2377-4155

傳　真／（02）2377-4355

E-mail／service@sanyau.com.tw

郵政劃撥／05844889 三友圖書有限公司

總經銷／大和書報圖書股份有限公司

地　址／新北市新莊區五工五路2號

電　話／(02) 8990-2588

傳　真／(02) 2299-7900

初　版／2013年8月

定　價／（精裝）新臺幣700元

　　　　（平裝）新臺幣500元

ISBN／978-986-6062-43-8（精裝）

　　　　978-986-6062-44-5（平裝）

http://www.ju-zi.com.tw
橘子&旗林 網路書店

國家圖書館出版品預行編目(CIP)資料

憶猶未盡之住家菜／劉冠麟 著. --初版--臺北市
橘子文化, 2013.08
面；　公分
ISBN 978-986-6062-43-8(精裝)
ISBN 978-986-6062-44-5(平裝)

1.食譜

427.1　　　　　　　　　　102010731

本書特別感謝 香格里拉台北遠東國際大飯店提供拍攝場地